D0350040

CLEAN
MEAT

CLEAN MEAT

How Growing Meat Without Animals

Will Revolutionize Dinner and the World

Paul Shapiro

With a foreword by Yuval Noah Harari,
author of *Sapiens*

GALLERY BOOKS

New York   London   Toronto   Sydney   New Delhi

Gallery Books
An Imprint of Simon & Schuster, Inc.
1230 Avenue of the Americas
New York, NY 10020

First Gallery Books hardcover edition January 2018

GALLERY BOOKS and colophon are registered trademarks of Simon & Schuster, Inc.

For information about special discounts for bulk purchases, please contact Simon & Schuster Special Sales at 1-866-506-1949 or business@simonandschuster.com.

The Simon & Schuster Speakers Bureau can bring authors to your live event. For more information or to book an event, contact the Simon & Schuster Speakers Bureau at 1-866-248-3049 or visit our website at www.simonspeakers.com.

Interior design by Jaime Putorti

Manufactured in the United States of America

10   9   8   7   6   5   4   3   2   1

Library of Congress Cataloging-in-Publication Data is available.

ISBN 978-1-5011-8908-1
ISBN 978-1-5011-8910-4 (ebook)

Dedicated to every person who sees

an apparently intractable problem and sets about laboring

tirelessly to solve it, confident in the words of Nelson Mandela:

"It always seems impossible until it's done."

CONTENTS

FOREWORD

By Yuval Noah Harari

Today, most of our planet's big animals live in industrial farms. We imagine that the earth is populated by lions, elephants, and penguins, roaming freely across vast savannas and oceans. That may be true of the National Geographic channel, Disney movies, and children's fairy tales, but it is no longer true of the real world outside the TV screen. The world contains forty thousand lions and one billion domesticated pigs; five hundred thousand elephants and 1.5 billion domesticated cows; fifty million penguins and fifty billion chickens. In 2009, a census counted 1.6 billion wild birds in Europe, including all species together. That same year, the European meat and egg industry raised close to seven billion chickens. A large portion of vertebrate animals living on our planet are no longer free-living but rather are owned and controlled by one animal: *Homo sapiens*.

These billions of animals are treated on industrial farms not as living creatures who can feel pain and distress but as machines for producing meat, milk, and eggs. They are often mass-produced in factory-like facilities, and their very bodies are shaped in accordance with industrial needs. The animals then pass their entire lives as cogs in a giant production line, and the length and quality of their existence is determined by the profits and losses of agribusiness corporations. Judged by the amount of suffering it causes, industrial farming of animals is arguably one of the worst crimes in history.

Up till now, scientific research and technological invention tended to worsen the lives of farm animals. In traditional societies such as ancient Egypt, the Roman Empire, or medieval China, humans had a very partial understanding of biochemistry, genetics, zoology, and epidemiology. Consequently, their manipulative powers were limited. In medieval villages, chickens ran free between the houses, pecked seeds and worms from the garbage heap, and built nests in the barn. If an ambitious peasant tried to lock a thousand chickens inside a crowded coop, a deadly bird-flu epidemic would probably have resulted, wiping out all the chickens, as well as many of the villagers. No priest, shaman, or witch doctor could have prevented it.

But once modern science deciphered the secrets of birds, viruses, and antibiotics, humans could begin to subject animals to extreme living conditions. With the help of vaccinations, medications, hormones, pesticides, central air-conditioning systems, automatic feeders, and lots of other novel gadgets, it is now possible to cram tens of thousands of chickens or other animals into tiny coops, and produce meat and eggs with unprecedented efficiency—but also with unprecedented misery.

In the twenty-first century, science and technology will give us even more power over our fellow beings. For four billion years, life on Earth was governed by natural selection. Soon it will be governed by human intelligent design. But technology is never deterministic. We can use the same technological breakthroughs to create very different kinds of societies and situations. For example, in the twentieth century, people could use the technology of the Industrial Revolution—

trains, electricity, radio, telephone—in order to create communist dictatorships, fascist regimes, or liberal democracies.

Similarly, in the twenty-first century, biotechnology could be used in many different ways. On the one hand, we could use it to design cows, pigs, and chickens who grow faster and produce more meat, without any thought about the suffering we inflict on these animals. On the other hand, we could use biotechnology to create clean meat—real meat that is grown from animal cells, without any need of raising and slaughtering entire creatures. If we follow that path, biotechnology may well be transformed from the nemesis of farm animals into their salvation. It could produce the meat so many humans crave without taking such an enormous toll on the planet, since growing meat is much more efficient than raising animals to later turn into that same meat.

Clean meat isn't science fiction. As you'll read in this book, the world's first cultured hamburger was produced and eaten in 2013. True, it cost $330,000, aided by funding from Google cofounder Sergey Brin. But we should remember that it cost billions of dollars to scan the first human genome, and today it costs just a few hundred dollars. In fact, by 2017—only four years after the first cultured burger tasting—the people who produced that first clean burger have already refined their process so that their costs are now a small fraction of that first burger. Competitors have already sprung up, including an American company that produced the world's first cultured meatball in 2016 at a comparatively bargain price of just $1,200. In 2017, the same company produced the first clean chicken sandwich and duck à l'orange, for an even lower cost, and they, too, intend to be marketing products in the fairly near future. With proper research and investment, within a decade or two we could produce clean meat on an industrial scale, which will be cheaper than raising cows and chickens. If you want a steak, you could just grow a steak, instead of raising and slaughtering an entire cow.

The transformative nature of this technology is hard to over-state. Once the price of clean meat is low enough, it will make not only ethical sense but also economic and ecological sense to replace slaughterhouse meat with clean meat. Animal farming is one of the chief causes of global warming, with the United Nations comparing animal agriculture's greenhouse gas emissions to those of the entire transportation sector. Even beyond climate, animal farming is one of the main consumers of antibiotics and poison, and one of the fore-most polluters of air, land, and sea. It may be easy to point our fingers at the oil and coal companies when lamenting the planetary prob-lems *Homo sapiens* are causing, but the conventional meat industry is a comparable polluter. Just as we need clean energy to replace fossil fuels, we need clean meat to replace factory farms. Switching to clean meat will be crucial for saving the planet from disastrous climate change and ecological degradation.

In this fascinating and hopeful book, Paul Shapiro highlights the great promise of this new method of food and clothing production—cellular agriculture. Thanks to this method, humans may soon stop raising and slaughtering farm animals by the billions. In the not-too-distant future, we may look back at industrial animal farming with the same horror that we today look back at slavery: a dark chapter in the history of humankind, which we have mercifully left behind us.

In the twenty-first century, technology will give us divine abilities of creation and destruction. But technology will not tell us what to do with it. When we come to design this brave new world, we should take into account the welfare of all sentient beings, and not just of *Homo sapiens*. We could use the wonders of bioengineering to con-struct either paradise or hell. It is up to all of us to choose wisely.

CLEAN MEAT

1

THE SECOND DOMESTICATION

On a muggy August day in 2014, I found myself wandering through the Brooklyn Army Terminal, a former World War II–era train station in New York's hippest borough which is now home to several dozen start-ups. Train cars from two generations ago stood motionless on their tracks, surrounded by brand-new, and mostly empty, office spaces. Could this place, frozen in time, really be the headquarters of a biotech company that, along with several other start-ups, is currently pioneering a technology that promises to upend our current food system?

As someone who's devoted my career to making our agricultural system more sustainable, especially through my work at the Humane Society of the United States, I've visited many food start-ups that claim their products will save the planet and prevent many of the illnesses that plague us—all while providing enough food to feed the world's growing population. Yet almost invariably they're located in the Bay Area, close to the Silicon Valley wealth that created them and continues to drive them toward the better future they're seeking to create. To me, Brooklyn seems far more bearded hipster than bio-tech haven, but this is where Andras Forgacs invited me to visit his new company, Modern Meadow.

Neither "modern" nor "meadow" came to mind as I surveyed the surroundings. The former military supply station was bought by New York City in the early 1980s and has since been converted to of-fice space. Now home to several dozen tenants, the station contains mostly start-ups. One of them, Modern Meadow, is generating head-lines around the world.

After fifteen minutes of searching the terminal and passing sev-eral other biotech start-ups, I finally found the lab's entrance. Forgacs, in his late thirties, welcomed me into the humble-yet-pristine space with a warm smile. Just about a dozen employees worked for him at the time. I wondered if I was really about to see history in the making.

After I entered, Forgacs and I chatted about the Modern Meadow process: culturing cow cells to grow beef and leather outside of the bovine. In other words, producing genuine leather without having to slaughter the cow from whom it came. The company was founded in 2011 as the first commercial venture to grow meat and leather in a lab, and I'd read that Forgacs could (theoretically) grow the entire world's beef supply from just one microscopic cell. The implications of this technology, if it can be perfected and scaled, are of course tre-mendous, potentially allowing us to continue to eat and wear animal products without causing the suffering, waste, and environmental damage wreaked by our current agricultural system.

Although Modern Meadow was the first company founded to commercialize these products, Forgacs isn't alone in his efforts. Sev-eral other companies—including all those that will be profiled in this book—have since been founded with the goal of bringing cultured-animal products into the mainstream.

We toured the quietly humming reactors where the culturing happens, and then Forgacs shocked me with a simple question.

"Want to try a sample?"

I came expecting to see, not to eat, and after more than two decades of happily enjoying a vegan diet, the thought of consuming beef wasn't exactly appealing.

I was also aware that, at that time, far more humans had gone into space than had eaten meat grown in a lab. Until Modern Meadow's existence, only a few academics had ever actually cultured meat in vitro, and perhaps less than a couple dozen people the world over had consumed it—ever.

"I've not eaten meat in a very, very long time, so I'm not sure my review would be that valid," I managed to mutter, half joking and fully hoping I'd successfully talked myself out of the situation.

I also contemplated the cost of the food, knowing from news reports that any amount of this beef would have to be worth a fortune.

"Didn't the burger they just served up in Europe cost $330,000, from cell to bun?" I was referring to the now-famous first-ever lab-grown hamburger—funded by Google cofounder Sergey Brin—that had been cooked and eaten at a press conference in London just a year earlier.

"Don't worry," Forgacs assured me. "You're our guest. And it's only a small sample—a steak chip, if you will. Really, it only cost about one hundred dollars to produce. And that'll come way down soon."

I'd certainly eaten a lot of steak fries in my life, but a steak chip was another beast altogether. Forgacs didn't simply want to create cultured versions of foods we already enjoy, like burgers; he also wanted to invent entirely new culinary experiences. The idea for the steak chip—think of them as potato chips made of meat—came from the realization of how much cheaper it would be to make thin sheets of meat than it would be to grow more complex pieces. Just as someone might grab a stick of beef jerky at a gas station for a quick snack, might they also try a bag of steak chips? "High in protein, low in fat, and superconvenient. I'd want that," Forgacs offered with a grin.

Initially on the fence, I quickly recognized the opportunity to be one of the first people ever to try a food that was generating so much buzz—and controversy—that I decided to accept my host's generous offering.

Forgacs pulled the steak chip from its container. I smiled and held it, wondering how my body would react to its first bite of meat in more than twenty years. I had little ethical concern about eating the meat, but it still felt bizarre to be on the precipice of ingesting animal flesh, especially flesh as novel as this.

I didn't decide to take a permanent vacation from meat because I didn't like eating it; I always enjoyed it as a kid and still enjoy the plant-based meats that are increasingly popular among omnivores today. Rather, I became vegan in 1993 having learned as a young teenager about the consequences of a meat-centric diet. Humans don't need to eat animals in order to be healthy, and the meat industry causes a lot of problems for animal welfare and the planet. So I figured why not do what I can to reduce this harm by leaving animals off my plate? Eating lower on the food chain also allows more food to be produced, since so many resources—like grains and water—are needed to feed livestock. Such efficiency has become even more important as the global population continues to boom.

Eventually, my love of animals led me to a career in animal protection, helping to spearhead legislative and corporate campaigns both to gain protections for farm animals and reduce the number of them being raised and slaughtered for food in the first place by helping people enjoy more plant-based meals. I'd been reading and talking about the concept of growing meat in a lab for years and always thought it was a promising solution to a vexing problem but never thought of the theoretical food as a product for myself as much as it was for those who were wedded to meat.

Yet here I was about to add real animal meat—albeit a slaughter-free version of it—back into my diet, at least for the day. The chip looked like a thin piece of jerky. As I stared at it, I contemplated just how remarkable—technologically and symbolically—this little piece of dried beef was. Perhaps I was holding in my hand the answer to so many of the problems animal agribusiness poses for humanity and the planet that hosts us. I raised the meat to my mouth, took a breath, and placed it on my tongue.

I've read accounts of other longtime vegetarians who've experienced all types of sensations after tasting meat for the first time in years: everything from a rush of endorphins and euphoria to nausea, stomach pains, and vomiting. But nothing like that happened to me. I chewed the steak chip, it tasted good, and it reminded me of barbecue.

My mind raced with questions: Was I about to get sick? Was I still a vegetarian? Did that even matter?

In actuality, it doesn't really matter if vegetarians or vegans will eat meat that was grown rather than slaughtered. They're not the intended audience. The real question—the one running through my mind in the Modern Meadow office and the one that's a subject of this book—is whether meat-eaters will accept this new method of producing the beef, chicken, pork—and a whole host of other animal products—that have come to form such a substantial part of our diets. Would we, as a society, at least consider easing ourselves into lab-grown animal products by first wearing some of Modern Meadow's in vitro leather? (The company is now focused exclusively on growing leather while others tackle meat.) And even if we'll accept such foods and clothing, can Modern Meadow and other culturing companies bring their products to market in time to correct the damage currently being inflicted by animal agriculture? In short, was that modest, albeit pricey, steak chip a preview of the future of food?

—————————

Our species is facing a crisis: as the global population swells, just how are we going to feed billions more people on a planet already suffering from a shortage of natural resources? Humanity's population has doubled since 1960, but our consumption of animal products has risen fivefold, and it's projected by the United Nations to keep rising. Complicating matters further, as poorer nations like China and India (which are also the most populous in the world) become richer, many of their citizens who'd previously subsisted on a largely plant-based diet will start to demand a more conventionally American regimen, heavy in meat, eggs, and dairy—products previously reserved for the wealthy but which they can now afford. As many sustainability experts observe, given how inefficient it is to raise animals rather than plants for our food, the earth just can't accommodate such an increase in animal-product demand. The change in climate will be too great, the deforestation too severe, the water use too massive, and the animal cruelty too unbearable.

Projections show that by 2050 there'll be nine to ten billion humans walking the earth. If most of them have the means to eat as lavishly as Westerners—particularly Americans—do today, it's hard to see how we can support the massive amount of land and other resources that will be needed to satisfy this demand. For the American palate alone, more than nine *billion* animals are raised and slaughtered for food annually, not counting aquatic animals like fish, who are counted in pounds, not as individual animals. In other words, more animals are used for food in America in just one year than there are people on the planet. And nearly all of those animals are confined for life inside factories that more closely resemble gulags than farms.

The green revolution—in which agricultural research led to huge increases in crop yields—dramatically expanded humanity's ability to produce more food with fewer resources, but the time we bought ourselves when we increased our agricultural productivity is running short, and we need to innovate our way out of the new agricultural crisis of our own making.

To put the problem in perspective, imagine walking through the poultry aisle of your local supermarket. For each chicken you see, envision more than one thousand single-gallon jugs of water sitting next to it. Then imagine systematically, one by one, twisting the cap off each jug and pouring them all down the drain. That's about how much water it takes to bring a single chicken from shell to shelf. In other words, you'd save more water skipping one family chicken dinner than by skipping six months of showers.

California and other drought-stricken areas may be content for now to impose restrictions on lawn care or suggest shortening shower times, but as the demand for water grows increasingly intense, no amount of individual restraint can make up for the amount of water required to sustain—not to mention grow—our animal-agriculture system.

And it's not just chicken.

It'll be increasingly hard to ignore the fifty gallons of water behind every single egg, easily enough to fill your bathtub to the brim. Or the nine hundred gallons of water needed for every gallon of cow's milk (now you're talking about a few hot tubs' worth of water). By comparison, you save eight hundred and fifty gallons of water when you buy a gallon of soymilk instead of cow's milk.

These stark inefficiencies remain regardless of whether we're talking about local, organic, non-GMO, or other buzzwords often labeled on animal-product packaging. Such facts make it clearer than ever that, as our population grows, if we're going to continue consuming meat, milk, and eggs at anywhere near the quantity we do today, we'll need to get more efficient—*much* more efficient.

Today, a group of scientists and entrepreneurs is trying to accomplish just that. Their goal: to grow *real* meat so that omnivores can continue enjoying beef, chicken, fish, and pork without having to raise and slaughter animals. If these start-ups succeed, they may do more to upend our dysfunctional food system than perhaps any other innovation, while addressing many of the biggest problems we face—

from environmental destruction and animal suffering to food-borne illness and perhaps even heart disease. These young companies are racing to make real a world in which we can have our meat and eat it, too: where we can enjoy abundant amounts of meat and other animal products without all the environmental, animal welfare, and public health costs.

Forgacs and his team at Modern Meadow are hardly the first people to think about growing animal products without raising entire animals. In addition to the imaginations of many sci-fi writers (perhaps most prominently Margaret Atwood in her novel *Oryx and Crake*, and even earlier in *Star Trek*), plenty of forward thinkers—including several outside of science or science fiction—have predicted that such a shift is inevitable. One of them would even become one of the most important figures in Western history.

"We shall escape the absurdity of growing a whole chicken in order to eat the breast or wing, by growing these parts separately under a suitable medium," proclaimed Winston Churchill in a 1931 essay entitled "Fifty Years Hence." His predicted time frame was admittedly a few decades off, but his foresight was remarkable, essentially foreshadowing the technology that would make Modern Meadow and its steak chips possible. "The new foods will from the outset be practically indistinguishable from the natural products," the future prime minister continued, "and any changes will be so gradual as to escape observation."

Churchill was predicting a major disruption in the way humans, for millennia, have obtained our protein. Not unlike the way cars largely relegated horse-powered travel to the history books, he was anticipating a technological advancement, he believed, would completely transform our relationship with a whole category of animals. Nor was Churchill the first to make such a prediction. As early as 1894, then-famed French chemistry professor Pierre-Eugène-Marcellin Berthelot claimed that by the year 2000, humans would dine on meat grown in a lab rather than from slaughtered animals. When pressed by a reporter about the feasibility of such meat production, Berthelot replied, "Why not, if it proves cheaper and better to make the same materials than to grow them?" Like Churchill, Berthelot's timing was off, but perhaps not by much.

Humans have always sought ways to improve what we eat. For most of our existence, *Homo sapiens* subsisted by foraging and hunting. Then, ten thousand years ago, some of us began shifting from the spear to the seed, domesticating plants and later animals in a veritable agricultural revolution. We soon began culturing, starting with products like beer and yogurt—perhaps the first biotech foods. And within the last century, the industrialization of our food supply revolutionized our options yet again, enabling immensely greater yields that can support—and encourage—an ever-increasing population explosion.

Today we may be witnessing the start of the next food revolution: cellular agriculture, the process of growing foods—such as real animal meat and other animal products—in a lab while leaving the animals alone and perhaps returning huge swaths of cropland back to their more natural habitats. Using technology first developed by academics and the medical field and now being commercialized by several start-ups, innovators are taking tiny biopsies of animals' muscle and then culturing those cells to grow more muscle outside the animals' bodies. Some entrepreneurs are even ditching the initial animal starter cells altogether and are growing—from the molecule up—real milk, eggs, leather, and gelatin that are all essentially identical to the animal products we know—even though they never involved a living animal at all.

With this new application of the technology, the start-ups you'll meet in this book are working hard to make Churchill's vision come to life. Right now, as I write this, these companies are producing real animal products from microscopic animal cells—or even from yeast, bacteria, or algae—that have the potential to revolutionize the food and fashion industries as we know them. At the same time, they offer the promise of solving enormous environmental and economic challenges posed by our growing global population—that is, assuming they can get the funding, regulatory approval, and consumer acceptance necessary to market their products on a global scale.

Unlike the also-promising plant-based protein revolution already well under way—the one that's given us brands like Tofurky, Silk soy milk, and Beyond Meat—these lab-grown products are not alternatives to meat, milk, and eggs; they are real animal products. Such technology may seem entirely novel, but in fact, nearly every bite of hard cheese you eat today contains rennet—an enzyme complex that causes milk to curdle and that, traditionally, had to be extracted from a calf's intestines—that was synthetically produced via nearly identical processes to those used by many companies in this book. And if you're a diabetic, you're almost certainly regularly injecting yourself with human insulin that was produced via the very same biotech process, too.

Meanwhile labs have, for years, been using similar processes to create real human tissues for experimentation and transplantation purposes. For example, a lab can take a patient's skin cells, culture them to grow new skin, and create real human skin that's identical to the skin the patient was born with. The body doesn't seem to know the difference because there is no difference—other than the fact that it was grown outside the body.

By applying what have largely been medical technologies to growing animal-agricultural products, scientists are developing what Dr. Uma Valeti, CEO of the cellular-ag start-up Memphis Meats, calls the "second domestication."

In the first domestication, thousands of years ago, humans began selectively breeding livestock and planting seeds and were therefore able to exert more control over where, how, and in what quantities we produced food. Today, we're taking that control down to the cellular level. "It's a clean-meat process," Valeti says, "that will let us produce meats directly from high-quality animal cells, thus using only the best quality muscle cells to produce the best meats." One of Valeti's investors in Memphis Meats, Seth Bannon, likes the analogy. His venture capital fund—named Fifty Years, in a tip of the hat to Churchill's essay—exists to help founders like Valeti. "Traditionally, we've domesticated animals to harvest their cells for food or drink," Bannon says of the work Memphis Meats is doing. "Now we're starting to domesticate cells themselves."

The scientists and entrepreneurs in this book are seeking to correct the course of an animal-agricultural system that's at the center of so many global ills. They've started in different places and have different values, but their goal is the same: racing to make real their vision of a world in which we produce our meat and other animal products not through raising and slaughtering chickens, turkeys, pigs, fish, and cows, but rather through culturing processes that essentially remove living, feeling animals from the process altogether.

"I see beer being brewed or yogurt—the yeast and lactobacillus do not cry out because it's in a tank," Forgacs quipped to a journalist a year after our meeting at the Modern Meadow headquarters. "That's our goal, to apply those principles to an animal product. We don't have to industrialize sentient beings."

If these companies succeed, the potential benefits to the planet, animals, and our health are obvious. And of course, what's also obvious to the investors pouring tens of millions of dollars into these start-ups is that where there's a major disruption to be had, there are also riches to be made. In a December 2016 interview on CNBC, Bill Gates talked about the promise of these start-ups in a discussion about the new Breakthrough Energy Ventures fund that he and fellow billionaires, like Jeff Bezos and Richard Branson, created. "We'll have several dozen companies that we're going after," the Microsoft founder commented. "Even in areas like agriculture we'll have artificial meat, where there's already some people doing things there. That's a big source of emissions there. . . . And if you can make meat another way, you avoid a lot of the issues like cruelty, and you should be able to make a product that costs less money."

While Gates had been funding plant-based meat alternatives for years, in August 2017 he, along with fellow business titans like Branson and former General Electric CEO Jack Welch, began pumping investment into the clean-meat space. Branson enthusiastically celebrated the funding he and his colleagues provided one start-up, prophesizing that "I believe that in 30 years or so we will no longer need to kill any animals and that all meat will either be clean or plant-based, taste the same, and also be much healthier for everyone. One day we will look back and think how archaic our grandparents were in killing animals for food."

Just in terms of food safety alone, these products could be game-changing. In slaughter plants, there's great risk of fecal contamination, whether it's from feces on the animals—they often defecate when introduced to the novel and daunting environment of a slaughter facility—or from feces in the gut that can contaminate the meat during the slaughter and butchering process. Many of the most dangerous food-borne pathogens are intestinal bugs like *E. coli* and *Salmonella* that result from such contamination. Of course, with meat cultured outside the animal, there's no fecal matter to worry about; it's produced in a completely sterile environment. As we'll see later, this is a primary reason the Good Food Institute (GFI), which promotes cellular agriculture products, popularized the term "clean meat."

That's why at least some food safety advocates are cheering the advent of such clean meat. Michael Jacobson, PhD, the founder of the Center for Science in the Public Interest, is one of them. The man who's crusaded against the dangers of food additives like trans fats and olestra is optimistic about cellular agriculture. "It's a good way to have animal products that would be a lot safer to consume and more sustainable to produce," he tells me. "I'd be happy to eat it."

In addition to the food-safety benefits, growing our meat rather than raising farm animals would also dramatically reduce our risk of the kind of global pandemic that keeps public health professionals up at night. Bird flu outbreaks, especially in Asia, tend to kill millions of animals annually. But the big concern is that avian influenza could jump species into humans, which is exactly what caused the massive Spanish flu outbreak of 1918, infecting nearly a third of humanity and killing upward of fifty million people. And this was a time when the world's population totaled only 1.2 billion people, a fraction of the 7.5 billion who call Earth home only a century later. And as the population has risen, so has our mobility, with millions of people traveling around the world each day. If an outbreak on the scale of the 1918 pandemic occurred today, it could prove even more devastating.

In 2007, the journal of the American Public Health Association editorialized on the pandemic threats posed by chicken factory farms, observing:

> It is curious, therefore, that changing the way humans treat animals—most basically, ceasing to eat them or, at the very least, radically limiting the quantity of them that are eaten—is largely off the radar as a significant preventive measure. Such a change, if sufficiently adopted or imposed, could still reduce the chances of the much-feared influenza epidemic. It would be even more likely to prevent unknown future diseases that, in the absence of this change, may result from farming animals intensively and from killing them for food. Yet humanity does not consider this option.

A decade later, so far, humanity still doesn't seem to have considered the option the American Public Health Association suggested: drastically slashing animal agribusiness to cut down on the risk of a pandemic catastrophe. But even if the chance of such an event is low at any given time, there are even more compelling reasons in the near term to think about raising fewer animals for food.

A major pandemic could be catastrophic to our civilization, but the likelihood of it happening in any given year is minimal. Yet some of the threats that factory farming of animals pose are already manifesting themselves today. Perhaps most notably, we're now facing a crisis of antibiotic resistance in human medicine, a problem that many medical and public health professionals say is due to animal agriculture. About 80 percent of all antibiotics in America are fed to farm animals, not to treat illness but subtherapeutically as a means of promoting growth and preventing sickness in such overcrowded conditions. Concerned about the ability to continue using literally life-saving antibiotics in human medicine, the American Medical Association now calls for a federal ban on using antibiotics to promote growth in farm animals, but due to the ag and pharma lobby interests, the doctors' call so far has fallen on deaf federal ears.

On a planet where demand for meat is only increasing as more developing countries rise out of poverty, we know that the earth's finite resources simply won't allow other nations to gorge themselves on the meat-heavy diet that Americans and Europeans have come to expect. Historically, richer nations have been able to afford high levels of meat consumption while the poor have subsisted largely on grains, beans, and vegetables, with meat being regarded more as a less frequent treat.

Even though Americans have begun consuming somewhat less meat in recent years, as household income rises in countries like India and China, so is the demand for meat. Alarmingly, just as one example, China's per capita meat consumption has skyrocketed fivefold in the past three decades. Where once beef was referred to as the "millionaire's meat" in the nation, it's now a daily part of the diet for hundreds of millions of Chinese citizens.

At least since the publication of Frances Moore Lappé's *Diet for a Small Planet* in 1971, it's been clear that Earth isn't big enough to sustain a global population of American-style meat-eaters. "Imagine sitting down to an eight-ounce steak, and then, imagine the room filled with 45 to 50 people with empty bowls," Lappé wrote. "For the feed cost of your steak, each of their bowls could be filled with a cup of cooked cereal grains."

While externalized costs of production in the United States help make animal products artificially inexpensive at the cash register, producing meat is an exceedingly pricey way to feed ourselves. Even long before Lappé's seminal work, President Harry Truman urged Americans to cut back on meat (including poultry) and egg consumption by cutting out animal protein on Tuesdays and Thursdays to save resources for the postwar European-rebuilding effort.

Fast-forward to today, and the message is still clear. "The reality is that it takes massive amounts of land, water, fertilizer, oil, and other resources to produce meat," says global relief charity Oxfam, "significantly more than it requires to grow other nutritious and delicious kinds of food."

The biggest cost associated with raising animals for food is the feed they're given, and they need a lot. When you think of soy, you might think about tofu or soy milk, but the lion's share of soy grown in the world is used as animal feed, and those soybeans take up a huge amount of land. Sadly, animal feed is the leading cause of rainforest deforestation, essentially killing the lungs of our planet. The World Wildlife Fund points out this fact, observing that "the expansion of soy to feed the world's growing demand for meat often contributes to deforestation and the loss of other valuable ecosystems in Latin America." In other words, slogans like "Save the Rainforest" might be more instructive if they concluded with "Eat Less Meat."

The Center for Biological Diversity recognizes this crucial connection between what we put on our plate and whether many species will have a planet to live on. That's why the environmental nonprofit launched a campaign called "Take Extinction Off Your Plate," in which it urges eco-minded consumers to prevent the eradication of wildlife by taking action every time they sit down to eat. The sole recommendation of the anti-extinction campaign: "The planet and its wildlife need us to reduce our meat consumption."

The strain that high levels of meat production put on our planet becomes even more apparent when you consider climate change. "Preventing catastrophic warming is dependent on tackling meat and dairy consumption, but the world is doing very little," warns the Britain-based Royal Institute of International Affairs, perhaps the most prestigious think tank in Europe. The institute, also known as Chatham House, notes that animal agriculture is a leading contributor to greenhouse gas emissions and that "it is unlikely global temperature rises can be kept below two degrees Celsius without a shift in global meat and dairy consumption."

The bottom line is that it's grossly inefficient to use resources to grow grains in order to feed farm animals simply so we can then eat those animals. And since nearly all farm animals in our country are grain-fed, we're essentially throwing massive amounts of food away when we opt to consume meat.

Even when you consider the most efficiently produced meat—chicken—it still pales in comparison to plant-based proteins. Chickens require so much grain that we have to feed them nine calories just to get one calorie back out, and again: that's the *most* efficient meat. A lot of those calories are used for biological processes we don't care that much about: growing beaks, breathing, digesting, and more.

We just want the meat, but to get it, we need to waste a lot of food. Bruce Friedrich, executive director of GFI, compares raising chickens for meat to taking nine plates of pasta and throwing them straight into the trash every time we wanted to enjoy just one dish of spaghetti. Few of us would do that, but the difference between that and buying meat may not be that substantial.

Yet despite all the evidence that producing meat is so inefficient, it's been a difficult proposition for those of us who've come to expect a high-meat diet to voluntarily opt for plants over animals. Many people simply love to eat meat. As I can attest, even at social events for vegetarians, the plant-based meats (veggie burgers, "chik'n" tenders, and more) are typically the most popular among guests, with lonely tubs of hummus and vegetables often remaining relatively untouched.

Despite decades of advocacy from vegetarian and animal protection organizations, the percentage of Americans who are vegetarians has hovered around 2 to 5 percent for the last thirty years. Yes, we've gone from around 220 pounds of beef, pork, and poultry per person in 2007 to 214 pounds in 2016, but even with this modest decrease, Americans still rank among the heaviest meat-eaters on the planet.

Plant-based protein pioneers like Pat Brown, CEO of Impossible Foods—purveyors of a very meat-like burger made from plants—are trying to help omnivores eat less meat without sacrificing taste. Even before Impossible Foods had a single product on the market, it had raised $182 million in investments from Google Ventures, Bill Gates, and more. Brown, a Stanford biology professor, argues that if we're going to significantly reduce, let alone reverse climate change, there's no way we're going to do it without a major reduction in the consumption of animal products. "Take every car, bus, truck, train, ship, airplane, rocket ship—all together," Brown says. "They produce less greenhouse emissions than the animal-agriculture industry."

But what if we could have our meat and eat it, too? What if we could enjoy actual animal products like meat and leather without the substantial environmental and ethical concerns associated with them today?

Andras Forgacs and his colleagues in the just-budding cultured-animal-products industry are committed to making this possibility a reality. The projected environmental benefits for their animal products are stark. For example, a 2011 study published in the journal *Environmental Science & Technology* by an Oxford University researcher, Hanna Tumisto, estimated that cultured beef could require up to 45 percent less energy, 99 percent less land, and 96 percent less water than conventional beef. Admittedly, any life cycle analysis performed so early has limitations, since it's still unclear what technologies will be invented that will actually make cellular-ag products commercially viable. But it's likely that growing animal products rather than raising animals would be tremendously more resource-efficient. That's why a 2015 study published in the *Journal of Integrative Agriculture* comparing the environmental impacts of cultured meat in China concluded that "replacing meat with cultured meat would substantially reduce greenhouse gas emissions and the demand for agricultural land."

And the Chinese government does indeed seem interested. In September 2017, the China Science and Technology Daily, a state-run newspaper, covered one American company's efforts to bring clean meat to the People's Republic, provocatively asking readers to imagine a world in which "you have two identical products; one . . . you have to slaughter the cattle to get. 'The other' is exactly the same, and cheaper, no greenhouse gas emissions, no animal slaughter, which one would you choose?"

Hoping to provide those kinds of choices is a whole cadre of start-ups seeking to make that happen. Those companies—like Modern Meadow, Hampton Creek, Memphis Meats, Mosa Meat, Finless Foods,

SuperMeat, Future Meat Technologies, Perfect Day, Clara Foods, Bolt Threads, VitroLabs, Spiber, Geltor, and others—are seeking to disrupt and ultimately revolutionize our food and fashion industries, something their wealthy venture capital backers are banking on. As former Morgan Stanley senior vice president and *Forbes* writer Michael Rowland told me, "Cultured-meat technology, once perfected, will totally reshape our global meat supply. Our meat will be made with science, not animals."

That's exactly why so many environmental and animal welfare advocates are championing these companies. They see cellular agriculture as being akin to the clean energy movement. "Factory farming is kind of like coal mining," explains Isha Datar, CEO of New Harvest, a nonprofit devoted to advancing cultured-animal-product technologies. "It pollutes and it's damaging our planet, but it gets the job done. And cellular agriculture is like renewable energy when it was still in its nascence. It has the promise of getting the same job done, but without so many terrible side effects."

Knowing the potential size of the impact they could have, the women and men in the cellular-ag community are almost boundlessly optimistic about what they may be able to do with this new application of technology. And, interestingly, for the most part they don't consider one another rivals as much as they do friendly competitors, all working toward the same goal of producing meat and other animal products through culturing processes that may one day render humanity less reliant on the exploitation of chickens, turkeys, pigs, fish, and cows.

In attempting to commercialize a technology that most consumers have, at this point, barely even heard of, each of these companies has different ideas about how best to introduce it to the market. In the meat space, should we start with growing beef, considering that current beef production wreaks more havoc on the environment than

any other meat? Or should we start with chicken, since more chickens are slaughtered than any other animal (save perhaps fish)? Or should we begin with milk, which is simpler to produce? Or should we set food aside altogether and focus instead on producing lab-grown leather, since leather produced in the lab will probably be easier for the public to "digest" than cultured meat?

There's a ways to go before these companies' dreams are realized. At this point, it's likely that clean animal products will be on the market in a limited way in the very near-term future, but producing commodity clean meats that can compete on price is still years away. As Jason Matheny, one of the pioneers of the early cultured-meat movement, joked to me several years ago, no matter what year someone asks him how long it'll be until cultured meat is available for sale in supermarkets, his answer is always the same: "Perhaps five years." But because of the work described in this book, that window appears to be closing rapidly.

There are several significant barriers facing the entrepreneurs featured in this book, each of which they must overcome to have any impact at all. The first is simply bringing the cost down—way down. All of them believe they can do this (or else they wouldn't even be doing their work), but that belief is based on their faith that they'll make technological breakthroughs that have yet to be made.

They also want people to understand the barriers they're trying to overcome. Before they can persuade consumers to try one of their burgers, they still need to figure out just how to actually produce their meats at scale. Since much of the technology they're using was invented for medical, not food, purposes, the size and cost of what they can do are both quite limiting. For example, they'll need to find better scaffolds—the "bones" on which the muscle grows—since the scaffolds cultured meat is currently grown on are expensive and incapable of producing anything other than ground meat. (So meatballs and hamburgers but not chicken breasts or steaks.) They also need

to invent industrial scale bioreactors (aka fermenters) in which the muscle could actually grow at commercial scale—something that doesn't yet exist, since, at present, such reactors are typically only used for medical purposes.

Another barrier that could hinder their success, even if they could scale up and compete on cost, will be potential government regulations and other bureaucratic obstacles that could stall pathways to the market. We've been applying modern biotechnology to food for decades, but regulating agencies might still be skeptical of this particular technology, given how novel it intuitively seems, which could slow down the approval process.

And finally, there's the ever-important question of whether consumers will even want to eat these foods—no matter how high-quality or inexpensive they become. With a growing number of consumers demanding "natural" and minimally processed foods, will there be a backlash against cultured-animal products as so-called Frankenfoods, a name that some have used to describe genetically modified organisms (GMOs)?

But unlike ag giants such as Monsanto and Dow AgroSciences, which have been quietly introducing GMOs into the marketplace in recent decades, the relatively tiny start-ups creating cellular-ag products want the public to know exactly how they're making their meat. "Radical transparency" is the buzzword these companies regularly tout, constantly seeking to tell the story of exactly what they're doing and how they're doing it. They're convinced that if consumers understand that what they're doing isn't that different from some of the foods or medical interventions we already routinely use, the acceptance and even eagerness will be there.

Certainly some of the same voices that oppose the marriage of biotech and our food system have serious concerns about lab-grown animal products. Their arguments are laid out in later chapters. Interestingly, though, one of the most high-profile voices in the sustainable foods world, Michael Pollan, author of *The Omnivore's Dilemma* and *In Defense of Food*, is supportive of these entrepreneurs' work. "In general I do think all these efforts to find substitutes for meat are worthwhile, since one way or another we're going to need to reduce our consumption—for environmental, moral, and ethical reasons," Pollan told me in response to my question about his view of cultured meat. "What the workable substitute will look like is still unclear, but research into all options seems well worthwhile, given the magnitude of the problem."

Some of the groups and individuals that have led campaigns against GMOs don't necessarily share Pollan's open-mindedness on cellular agriculture. Many of them have legitimate concerns about the way some technologies have been used to create a less sustainable food system in the past. But technology is like a knife: it can be used to lovingly prepare food for friends, or it can be used to kill—it just depends on how it's applied. One thing that's certain, though, is that the success of clean meat would do an enormous amount to reduce the number of GMOs in existence. Right now, 90 percent of the GM crops planted in America are fed to farm animals. McKay Jenkins, a sharp critic of industrial agriculture, argues in his 2017 book *Food Fight: GMOs and the Future of the American Diet*:

> *The grand prize—growing meat from cell cultures rather than from actual living livestock—could mean all types of powerful changes to industrial agriculture. We wouldn't need pesticide-laden GM corn, industrial slaughterhouses, or gasoline, because we wouldn't be feeding, slaughtering, or shipping animals around the country. We also wouldn't need to deal with the mountains (or lakes) of animal waste that contaminate our water, or clouds of methane that contribute to climate change. And we wouldn't need to kill billions of animals to satisfy our bottomless desire for protein.*

Wherever the debate goes with regard to biotech and food, most of these companies want to portray their forthcoming products as natural and not dissimilar to many foods we already consume daily. Others, though, seem to embrace such characterizations of their foods as novel and foreign. One company, calling itself Real Vegan Cheese, isn't limiting itself just to making cow's milk cheese; it promises to make cheese from the (synthesized) milk of narwhal whales in order to "raise awareness of ocean health" and "show that the process will work with the genes of any sequenced mammal." Another, as you'll read in chapter 7, has already produced gummy candy using mastodon gelatin. (Yes, you read that right: lab-grown gelatin of the North American behemoth we drove to extinction millennia ago.)

The race is on to bring the world's first cultured animal products to market. Start-ups are attracting millions of dollars from some of the biggest names in the venture capital world, all of whom are hoping to upend the way we've fed and clothed ourselves for millennia, but especially in the last half century, since factory farming expanded our access to animal products, albeit with countless externalities. And of course they aim to make a handsome profit at the same time.

Is it possible that just in the way we now have local breweries specializing in their own craft beers we'll soon have local meat breweries? Modern Meadow's Forgacs believes so. "A brewery is a bioreactor. It's where cell culture takes place. Instead of brewing beer, we could be brewing leather or meat. It's not hard to envision."

It may be that not only will we have the ability to culture meat in breweries (perhaps called carneries?), we might even have the ability to do it in our own homes. Just as it's fairly unremarkable today to have a bread-maker or ice cream–maker in our kitchens, one day we may have meat-makers, too.

Dr. Mark Post, another scientist working in this space, foreshadows a time when people like him will be "selling tea bags of stem cells from tuna, tigers, cows, pigs, or whatever meat you want, and from the comfort of your own kitchen, you could grow your own meat."

As someone who's now eaten cultured beef, poultry, fish, dairy, and even foie gras, and held cultured leather in my hands, I wrote this book as an exploration of the promise that this emerging industry holds. My career in animal welfare has brought me to the front lines of the seemingly never-ending battles between the meat industry and animal as well as environmental advocates. Yet maybe that battle could come to an end with both sides winning: people will still eat meat, but neither the planet nor animals will be harmed nearly as much in the process. It's possible that with the commercialization of the products described in this book, we'll soon start seeing animal protection organizations campaigning with slogans like "Eat Meat, Not Animals."

The world is getting more and more crowded, and hungrier for resource-intensive animal products every year. Shifting toward a more plant-based diet would help ameliorate much of this crisis, and it's important that the plant protein sector continue to grow. But our species—and others with whom we share the planet—can't rely on just one solution to such a large problem. Just like with renewable energy, we need many alternatives to the problem.

If cellular-ag companies succeed, it could be the biggest upheaval in how we produce food since the agricultural revolution some ten thousand years ago. And it could just be the answer to some of the most pressing problems humanity faces as we move deeper into the twenty-first century.

2

SCIENCE TO THE RESCUE

It's difficult to envision a world in which we're no longer so reliant on animals for meat. After all, with the exception of some relatively new plant-based protein products, animals have satiated our species' desire for meat since the dawn of *Homo sapiens* some two hundred thousand to three hundred thousand years ago. But when we consider how many other things we used to depend on animals for—things like clothing, tools, shelter, and transportation—we realize how, in the past few centuries alone, new technologies have allowed us to dramatically reduce our reliance on animals across the board.

Prior to the twentieth century, for example, societies around the globe were largely lit by a ubiquitous fuel source: whale oil. This generated a gigantic whaling industry, one made even bigger by the Industrial Revolution's demand for mechanical lubricants in all the new factories. And no nation was as obsessed—or successful—with whaling as America.

New Bedford, Massachusetts, became known as the "City That Lights the World," and there were enormous fortunes to be made from the New England whaling fleets that traversed the high seas in search of giant prey. Whaling played such an integral role in America's economy that both the British in the Revolutionary War and Confederates during the Civil War waged attacks on the United States' whaling fleets. Moreover, the whaling industry commanded outsized influence on the economic and political life of the nation. Long before the advent of the modern petroleum oil industry, the whale oil industry reigned supreme during both colonial and early republic days. "The value of the oil and bone brought back to port made whaling," writes Eric Jay Dolin in his whaling history tome, *Leviathan*, "by the middle of the nineteenth century, the third largest industry, after shoes and cotton, in Massachusetts, and according to one economic analysis, the fifth largest industry in the United States." (For reference, the fifth largest industry in America today—measured by contribution to GDP—is durable manufacturing, bigger than all retail trade, all construction, and even the federal government.)

Today, the United States—including New Bedford—still has a large number of boats solely used to seek out and shoot whales, though now the shooting is done with cameras, and the only place you're likely even to find a harpoon is in a museum. In the twenty-first century, the United States is a leader not in whale-killing, but in whale-watching.

So how did an industry so powerful—indeed one of the most potent lobbies in antebellum America—go from hegemony to irrelevance?

It'd be simple to construct a narrative about the suffering of animals and sustainability concerns, about how great men and women fought the good fight on behalf of whales and beat back the Goliath that was the whaling industry. Indeed, there were early concerns raised about the ethics of whaling, mostly related to the ruthless efficiency with which whalers massacred their prey. Such wars on whales were likely to deplete the oceans of them entirely, some warned.

In fact, an 1850 letter to the editor of the *Honolulu Friend* newspaper, signed by a bowhead whale, begged the persecutors of arctic whales for mercy. Lamenting how many of his kind had been "murdered in 'cold' blood," the cetacean "author" of the letter noted that whales had recently held a meeting "to consult respecting our safety, and in some way or another, if possible, to avert the doom that seems to await all of the whale genus throughout the world." He continued, "I write in behalf of my butchered and dying species. I appeal to the friends of the whole race of whales. Must we all be murdered . . . ? Must our race become extinct? Will no friends and allies arise and revenge our wrongs?"

The whales would soon get their wish, but not for the reasons enumerated by our precocious sea-dwelling letter writer. The whaling industry's downfall was just about to begin, bringing it from its greatest heights to near obscurity in a mere couple of decades.

A cartoon in an April 1861 issue of *Vanity Fair* paints the picture quite vividly: It shows a ballroom filled with celebrating whales, all dressed in black tie, some raising their glasses, others clinking them together in cheerful gaiety. Banners adorn the ballroom, one proclaiming, "We Wail No More for Our Blubber."

The reason these whales were celebrating their liberation is straightforward: they had an innovative entrepreneur to thank for their lives—Canadian geologist Abraham Gesner.

Today, Silicon Valley investors salivate at buzzwords like "disruption." Had venture capitalists known Gesner, his patent on kerosene would've had them falling over themselves to empty their wallets into his newly commercialized product.

Kerosene, which is derived from petroleum, offered a much better yet more affordable alternative to whale oil. In 1854, when Gesner commercialized kerosene, the US whaling fleet annually slaughtered more than eight thousand whales in the high seas around the globe. But in the ensuing years, as more Americans switched from whale oil to kerosene to light their homes, the country's whaling fleet, which had grown annually during the entire first half of the nineteenth century, began rapidly contracting. From a high of 735 boats in 1846, within just three decades, the nation's whaling fleet was down to just thirty-nine ships. (That limited whaling continued largely to supply the women's corset market with whalebones, though that was later phased out by the invention of spring steel in the early twentieth century.)

That's right: in just thirty years, the whaling industry was decimated, shrinking *95 percent*, largely, though not entirely, because a better, cheaper alternative arose and supplanted it. Thanks in substantial part to Gesner's innovation and subsequent oil discoveries, untold numbers of whales were spared from grisly deaths and perhaps extinction altogether. As Dolin writes, "the viscous black oil that gushed out of the earth provided a challenge that could not be circumvented, becoming so plentiful, so versatile, and so cheap that it quickly replaced whale oil in many of its applications." And in the spirit of creative destruction in a free market economy, the kerosene lamp industry got a taste of its own medicine when it was subsequently later rendered extinct by Thomas Edison's electric light bulb.

A similar story is found in our cities' streets, once dominated by the sounds of cracking whips and screaming men, both targeted at the unfortunate horses who labored to transport us and our goods through heat and cold, rain and snow.

The American animal welfare movement was largely started in the late 1860s by pioneers like Henry Bergh, who deplored seeing the open and flagrant abuse of equines on a daily basis and consequently founded the American Society for the Prevention of Cruelty to Animals (ASPCA) in 1866. Animal welfare crusaders like Bergh campaigned for all types of reforms: watering stations for horses, mandatory resting hours, Sabbath resting days for them, and more.

There were so many horses in New York City that as author Jeff

Stibel notes in his book *Breakpoint*, in 1880 a committee of experts commissioned by the US government was assembled to predict what the city would look like by 1980. Their unanimous prediction: New York City would cease to exist within one hundred years, buried under a pile of horse manure. They calculated that given the city's unsustainable population growth rate, it would need to increase its equine labor population from two hundred thousand to six million within one hundred years. Already the city was burdened by the problem that each horse deposited more than two dozen pounds of manure and over a gallon of urine onto the streets daily. A thirty-fold increase in the number of horses would render the city unlivable.

Yet in the end, what freed horses from labor in our streets and what saved New York City from literally drowning in horse poop wasn't humane sentiment nor environmental concern. Just as kerosene helped save the whales, internal combustion engines helped replace horses as our primary means of transport. It was an inventor's imagination, not a social movement's moral argument, which rescued horses. And it's not as if the public was clamoring for cars before they existed. As Henry Ford himself famously declared: "If I had asked people what they wanted, they would have said faster horses."

Even today we still use terms like "horsepower" to describe how powerful a car is, yet, thankfully for horses, they were long ago freed by an innovative technology that the nation—and world—rapidly embraced. Wayne Pacelle, CEO of the Humane Society of the United States, remarks in his book *The Humane Economy*, it "was primarily Henry Ford and not . . . ASPCA founder Henry Bergh who was at the wheel in dramatically reducing cruelty to horses in the nineteenth and early twentieth centuries."

The new auto industry created its fair share of jobs, but not without decimating jobs in other sectors along the way. With the demise of the horse-drawn carriage also came the fall of a whole host of supporting fields. From buggy whip producers to the hay growers providing the feed for all the horses, within just a couple of decades, long-standing industries were mere shells of their former selves.

These historical examples have weighed heavily on the minds of many social reformers in the more modern era. If you want to solve social problems in today's world, would you be better off going into traditional career paths like nonprofit work, policy, or politics, or will you have a greater chance to make an impact in the for-profit fields of technology, engineering, and entrepreneurship? There's no doubt to me the former are important (I have, after all, spent the bulk of my career as a policy advocate), but the fact of the matter is, as long as people demand real meat, the market is going to supply it, and globally speaking, demand for meat is only going up.

Is it possible that factory farms will one day seem as archaic to us as a whaling ship; a slaughter plant as antiquated as a horse-drawn carriage? This is exactly what those in the cellular-agriculture community hope, and it's why one young idealist, burdened by the knowledge of how unsustainable meat production is, decided to start a movement.

—————————

In 2002, Jason Matheny, a twenty-seven-year-old Johns Hopkins University public health graduate student, scored a job with the Avahan project, an effort by the Bill & Melinda Gates Foundation to reduce the rate of HIV in India. His task: to try to make HIV control programs more effective in order to avert substantial suffering and save lives.

Packing up his minimal belongings, Matheny set off to the subcontinent where he'd spend the next six months working in some of the poorest communities in the world. His time with Avahan was largely spent collecting data and crunching numbers, something that suited the young researcher's analytical mind well. Yet even as he witnessed what he calls "truly unbearable human suffering," he was also repeatedly stunned by the "shocking misery" so many animals were enduring in India, from mange-ridden street dogs to starving and ownerless cows who wandered the streets and often died from intestinal blockages after ingesting plastic bags.

The only silver lining, he figured, was that animal suffering in India was limited to strays, and not on the same scale as what he calls "an agribusiness system that sentences billions of American farm animals to lives that just aren't worth living." It was disturbing, but at least it wasn't systematized.

Or so he thought.

A couple months into his trip, Matheny visited a village outside of New Delhi. As the sun beat down, he sat inside a modest shack, interviewing a woman who'd lost her husband to HIV. Seeing the now-fatherless children and hearing their mother lament how she couldn't earn enough to feed them all on her own, Matheny wanted to break down.

"The heat was intense, and I could feel my grip on the pen getting looser from all my sweat," he recalls. "Her story made me so distraught, and my handwriting became less and less legible. I remember the thought creeping into my mind: Couldn't we get a breeze in here?"

At just that moment, as if he'd been able to summon nature itself, a gentle gust of wind entered through the shack's open door, giving Matheny a brief respite from his discomfort. The relief didn't last long. Almost immediately, he detected a strong, foul odor that had entered on the draft.

Sensing her interviewer's disgust, the widow grasped his hand. "I'm sorry," she offered, "that's just the chickens."

"Really?" he replied, his interest piqued. "That's a chicken's manure?"

"Well"—she looked down—"*a lot of chickens'* manure."

He asked if he could see these chickens, so the widow took Matheny outside and pointed to a long, windowless warehouse three hundred or so yards away. It was about as local a farm as it could get.

The structure looked just like an American factory farm. Giant fans whirled furiously at the end of the building, pumping an artificial wind into one side and taking noxious fumes out the other end. His host led him toward the structure, the stench worsening as they approached. After a short walk to the source of the odor, she opened the door, and Matheny was stunned by what he found inside.

A blanket of tens of thousands of white chickens carpeted the floor from wall to wall. There was so little space in between each bird that it was difficult to notice the brown floor underneath them comprised of litter and feces. Dim bulbs on the ceiling offered sufficient light to know these were birds, but Matheny had to strain to notice they were individuals.

As far as Matheny's eyes—already burning from fecal ammonia in the air—could see was a mass of animals. It didn't appear that there was room for a human to walk among the birds without trampling them, but his host hastily walked in, beckoning her guest into the warehouse as if nothing was out of the ordinary.

Birds scattered, piling up on top of one another to make way for their visitors. They were so bulky that many had difficulty even taking a few frantic steps out of the way before collapsing. One appeared as if she was suffering cardiac arrest after being trampled by other birds jockeying toward her for space.

Matheny was inside the warehouse for just a few minutes, but the experience left an indelible mark on him. "Here I thought industrial animal agribusiness was only in the developed world, yet these chickens were living proof otherwise."

That night, back inside his modest Delhi apartment, Matheny took advantage of the fact that electricity was reliably flowing for the evening and began poring over the United Nations' Food and Agriculture Organization's website. A vegetarian himself, he knew that India had a rich history of vegetarianism, but he was surprised to learn that

Indian meat consumption, especially of chickens, was skyrocketing in accordance with the nation's ascent out of the third world. The same was true in other hugely populated countries like China, which also had a history of relatively low meat consumption.

"It was like feeling the earthquake out in the ocean and knowing that the tsunami was soon to strike land," Matheny analogized. "It began dawning on me that even if we could curb meat demand in America and Europe, if we didn't stop this trend of more meat in the developing world, where nearly all population growth will come from in the coming decades, those gains would be overshadowed by huge amounts of disease, environmental harm, and animal suffering. That's what got me wondering if there might be some type of technological fix that could be applied to the problem."

Months later, back in the United States, Matheny continued wondering what, if anything, could be done to protect the planet from such a troublingly unsustainable predicament. A true believer in the power of technology to improve society, he regularly read websites devoted to the latest and greatest tech advancements. Later that year, one particular headline caught his attention: "An In Vitro Edible Muscle Protein Production System (MPPS)."

Between 1999 and 2002, the article explained, a group of New York researchers funded by NASA turned into reality what a handful of futurists had only fantasized about since Churchill's prediction about lab-grown meat three-quarters of a century earlier. Led by Morris Benjaminson of Touro College in New York City, they'd isolated muscle cells from a goldfish and grown them outside the animal's body. The method was simply to take segments of goldfish skeletal muscle and bathe them in various nutrients that in the body would cause muscle growth, and that's exactly what happened. The researchers did fry up the fish meat they'd cultured to see how it would cook and smell—they said it was similar to conventional fish—though none of them ate the results of their experiment, lacking Food and Drug Administration approval. "Their goal in this case was to allow astronauts to cultivate meat in space," Matheny remembers, "but I kept thinking as I was reading the article, 'In space? Why not do this on Earth?'"

He started scanning the scientific literature to see if he could find any articles about growing meat in a lab for terrestrially bound human consumption. After turning up nothing, Matheny emailed the authors of the NASA paper and other tissue engineers, asking why no one had written any scientific articles about mass-producing what he referred to at the time as "in vitro meat."

Most wrote back, with largely similar responses: *Why would you want to do that? If people want alternatives to meat, they could just eat soy burgers.*

As an avid consumer of those soy-based products himself, Matheny indeed hoped people would switch to these and other plant-based alternatives, but he knew that a problem as big as increased global meat consumption required more than just one possible solution. Just like there are now many renewable alternatives to fossil fuels (think solar, wind, geothermal, and more), could there be more than one alternative to factory farming of animals? For whatever reasons, despite the existence of affordable and nutritious vegetarian food, it seemed that whenever a population began escaping poverty, it also began adding more animals to its diet.

"Humans really love to eat meat," he says. "It's a hard habit for many to break. There were already resources going into promoting and improving plant-based meat alternatives, but no one was exploring whether investment in this other idea of growing real animal meat was a viable alternative to factory farming."

Matheny also noted that, at the time, despite increasing awareness about the ills of factory farming, American meat consumption was going up, not down. And, as noted in the previous chapter, while there's been some modest overall reduction in our consumption of meat, we are still a primarily meat-loving nation. In short, the problem was (and remains) so urgent and severe that we don't have time to wait for a massive dietary shift toward plants.

"You can spend your time trying to get people to turn their lights out more often," Matheny observes, "or you can invent a more efficient light bulb that uses far less energy even if you leave it on. What we need is an enormously more efficient way to get meat." Just as Gesner did with whale oil and Ford did with horses and buggies, Matheny wanted to render conventional meat obsolete by developing an alternative that still satisfies consumer demand for meat.

During World War II, Americans became accustomed to meat rationing as part of the effort to support the troops abroad. When the war ended, the country was obviously flying high after its victory over Germany and Japan, but the problem of meat availability remained.

Wartime government-induced price ceilings on meat meant that many farmers simply chose not to produce in fear of losing money, but Americans were willing to endure the less-meat diet out of a shared desire to defeat the Axis enemies. When the war ended, however, those ceilings were lifted, and unsurprisingly the price of meat skyrocketed. In the midst of the 1946 midterm elections, President Truman, trying to save the Democratic Party, moved to enforce price ceilings again, but this time without the ability to rely on patriotism to encourage farmers to produce meat for untenably low prices. Enraged, the meat lobby retaliated by having producers again cease sending animals to slaughter.

As *TIME* magazine's Emelyn Rude wrote of the incident in 2016, "Miners declared they could not work without more meat and began striking in Washington. Hospitals stirred scandal by claiming they could only find horsemeat to serve their sick patients. The lines outside the butcher shops still in operation stretched for blocks and provoked shoving and scratching amongst the patrons." The country was on edge. *TIME* itself even editorialized on the topic at the time, blaming what it called the "widespread meat famine" directly on Truman. (Keep in mind that what was considered a high-meat diet in America in the 1940s would be considered a reduced-meat diet today, since per capita consumption has risen every decade since then.)

Congressional Democrats begged the president to do something about the meat crisis, declaring it the single issue of concern to their constituents. The Republicans campaigned on a pro-meat message, throwing metaphorical red meat to voters who were deprived of the real thing. "Got enough meat?" asked Representative John Vorys, a Republican from Ohio in a campaign speech, a slogan which quickly came to be an election rallying cry. Representative Sam Rayburn even dubbed the midterm election of 1946 the "Beefsteak Election."

Blaming the meat barons of the time for the problem, Truman called them "the same group [that] hated Franklin D. Roosevelt and everything he stood for," and castigated them as a "reckless group of selfish men." But those men had played their hand well, and the meat shortage they caused by withholding animals from slaughter forced the president to relent, lifting all meat price ceilings. But it was too late: in substantial part due to the perception of a meat shortage, the Democrats lost control of both chambers of Congress.

This story, dramatic as it may seem today when many Americans are accustomed to having access to nearly unlimited amounts of any food we want whenever we want, illustrates just how strong the human desire for meat can be, and how hard it is to convince people to voluntarily reduce their meat consumption once they've developed a daily meat habit. And this isn't just true in America. Every culture that cultivates a taste for high meat consumption seems to favor it strongly. As Jason Matheny found when he was in India, when societies that've historically been too poor to sustain high rates of meat consumption start getting richer, the first thing they do is add more meat to their diet.

Even tribal people who enjoy nearly none of the benefits that modern prosperity has brought to developed countries, including high meat consumption, still often tie their own welfare to how often they can eat meat. *National Geographic* reports that the Tsimane Indian tribe in the Bolivian Amazon considers meat essential to their sense of well-being. "The children are sad when there is no meat," one mother told a journalist through an interpreter.

As the world continues to add more and more people, many of them in the developing world where demand for meat is growing at a fast clip, the question of how we're going to avoid "meat famines" like the one that affected American politics in 1946 is a pressing one.

As noted, projections show that by 2050, nine to ten billion humans will inhabit the planet. The problem is that as our population expands, our access to other planets' resources isn't expanding with it. The earth is just, to borrow a phrase from astronomer Carl Sagan, a pale blue dot in our solar system and we're exploiting it with alarming rapidity today, emptying its oceans of fish and razing its forests for cropland and pasture, mostly for animal agriculture.

Already today, according to the Food and Agriculture Organization of the United Nations, more than a quarter of the earth's ice-free land is used for livestock grazing, and a third of our cropland is devoted to feeding our farm animals. As *New Scientist* reported in 2017, "if we were all determined to avoid animal proteins, the majority of agricultural land would be released from production, freeing up huge areas for wildlife." On the other hand, if most of the billions of incoming people on earth expect to eat anything even close to what richer populations do today, just where are we going to produce all this meat?

The green revolution may have allowed our population to expand without global famine, but even its architect, Norman Borlaug, warned that the expanding population would render advancements like his hybridizing of wheat insufficient. In his 1970 Nobel Peace Prize acceptance speech, the man credited with saving a billion human lives from starvation ended on a somber but hopeful tone:

> *The green revolution has won a temporary success in man's war against hunger and deprivation; it has given man a breathing space. If fully implemented, the revolution can provide sufficient food for sustenance during the next three decades. But the frightening power of human reproduction must also be curbed; otherwise, the success of the green revolution will be ephemeral only. Most people still fail to comprehend the magnitude and menace of the "Population Monster." . . . Since man is potentially a rational being, however, I am confident that within the next two decades he will recognize the self-destructive course he steers along the road of irresponsible population growth and will adjust the growth rate to levels which will permit a decent standard of living for all mankind.*

Borlaug's confidence in humankind's rationality on this point so far seems largely unfounded. In the nearly fifty years since he made this speech, the human population has continued to rise and it shows little sign of abating in the coming couple decades. In the absence of our ability to slow or reverse this trend, now is the time to start thinking seriously about how we're going to feed our future selves. As Bruce Friedrich of the Good Food Institute wrote in *Wired* in 2016, "We're not going to feed the world, and we're not going to avoid a climate catastrophe, if we continue our global reliance on a system of food production that is so vastly inefficient and polluting. Individual change is important, but institutional change is even more important."

It was exactly this type of institutional change that Matheny hoped to inspire. After coming across the original NASA research about in vitro meat production in 2002, he continued corresponding with the scientists behind that study. After reading more and more literature about tissue engineering, he and a few of them became convinced that growing real meat from farm animals outside their bodies could indeed be done.

In the meantime, in 2003, an Australian artist named Oron Catts, along with a tissue engineering friend, Dr. Ionat Zurr, decided to grow some frog leg muscle in vitro and serve it up to diners in France as part of an art exhibit. While the tasters reportedly spat the meat out in disgust, the controversial exhibit garnered the headlines Catts desired, landing his project on Matheny's radar and only intensifying his interest. "The frog legs didn't seem that appetizing to me, but they definitely proved the point that you can do this with pretty much any animal," Matheny says.

Thinking about a beer brewery he'd once toured, Matheny daydreamed of beef breweries churning out clean, safe meat—all while freeing animals from factory farms and giving the earth a needed reprieve from the oncoming economic and environmental disaster that rapidly increasing global meat demand would ensure.

All he needed to do was spark enough interest to get money flowing to the right kind of research. Toward that end, in 2004, Matheny founded the first organization devoted to promoting research into growing real meat without animals. After conducting informal focus groups of a number of his friends in the sustainability and animal welfare fields, he settled on a name: New Harvest. "The name really encapsulated exactly what I was trying to bring about: a new type of bountiful harvest for humanity. One that would feed us with safe and nutritious food without destroying the earth in the process."

New Harvest's first task would be simply to get governments and other potential funders as excited about the prospect of lab-grown meat as Matheny was. His efforts to gain the attention of the US Department of Agriculture didn't go far, perhaps because the agency has long championed increased American farm animal production, or perhaps because in the absence of a single company in the lab-grown meat field, such research would seem too far removed from the agency's interests. For whatever reason, Matheny wasn't able to get an audience.

As a result, he started scanning other nations' governments for assistance. The European Union has been very skeptical of certain new food science applications, such as genetically modified crops, but the European Union did seem to have much more willingness to regulate its animal-agriculture sector than the United States. For years, concerns about the environment and animals had led the European Union to adopt reforms that might mean it would be open to eco-friendlier ways to produce protein. Matheny learned that the Netherlands, pressed by a number of committed environmentalists in its government, had for years been investigating alternative protein sources derived from plants rather than animals. In response, the Dutch government initiated a project, Protein Foods, Environment, Technology, and Society (PROFETAS), which championed pea protein production as an efficient protein of the future, in part since peas could be easily grown in the Netherlands.

After founding New Harvest, Matheny wrote PROFETAS asking why they didn't consider "in vitro meat." The leaders of the group were interested in Matheny's suggestion, especially since they must have been aware that an eccentric Dutch scientist, Willem van Eelen, had for years been trying, with only modest success, to culture meat. Matheny was familiar with Van Eelen's work and had, in fact, written to him several times without getting a response.

Born in Indonesia to Dutch parents, Van Eelen had been captured by the Japanese while serving in World War II. Living in a POW camp, he thought about food all the time—especially how to get the most out of the meager portions offered. Seeing emaciated dogs—their ribs visible, begging for scraps from hungry prisoners—took a toll on the inmate. Van Eelen fantasized about producing meat essentially from thin air so that no one would have to go hungry.

After the war, now living in Amsterdam, Van Eelen pursued a medical degree, during which he saw, as part of his education, an actual muscle gain mass outside the body. Since meat is primarily just muscle, why, he figured, couldn't we produce food that way? And so even while he practiced as a medical doctor, he spent decades tinkering, never working full-time on the project, trying to make muscle grow in vitro.

Finally, in 1999, Van Eelen persuaded the European Union to grant him a patent on a basic cultured meat production method. Part of what he was doing involved taking whole pieces of tissue from an animal and making them grow at the edges. Even though he never got the muscle to continue growing—there was a limit to the cell divisions possible—he'd succeeded in expanding its mass. (The patent was broader than this one process, and in fact was so potentially useful that in 2017 a new player to the clean meat space, Hampton Creek, purchased it, elating Van Eelen's daughter, Ira van Eelen, who maintains great hope that her father's dream will be actualized.)

Admittedly, Van Eelen had had a hard time persuading the Dutch government to fund his research, but Matheny wondered if he might have more success when the Dutch invited him to present at the 2004 PROFETAS conference in Wageningen. While there, the young American managed to get a private meeting with the Netherlands' minister of agriculture, where he presented the case for government funding of cultured-meat research. If the Dutch are serious about wanting to help protect the planet, Matheny argued, plant-based proteins are a good start, but the problem is too big to put all hope on them alone. It'd be like trying to move away from fossil fuels and putting all your research into just wind power, while ignoring the role other clean energy sources, like solar, might have to offer. The world needed re-search into lab-grown meat.

Months after he left, to his great surprise and delight, Matheny got word that his efforts had paid off: $2 million euros would soon be devoted to the experiments, which would be carried out at three Dutch universities.

The pledge of funds from the Dutch government was a huge step forward for New Harvest, energizing Matheny and causing him to begin seeking to rectify the total dearth of academic litera-ture on the topic. Citing the tissue engineering work the medical community had been pioneering for years, he persuaded some of that community's scientists to join him in writing a blueprint for how mass-cultured-meat production could actually work. And so it came to be that the first-ever scientific article outlining just how one could produce cultured meat was written.

"*In Vitro*–Cultured Meat Production," was published in the jour-nal *Tissue Engineering* in 2005. In the paper, three tissue-engineering researchers—Peter Edelman, Doug McFarland, and Vladimir Mironov—joined Jason Matheny in laying out the case for the po-tential of this new technology. The scientists explained that tissue-engineering technologies used in biomedicine could more easily succeed in cultured-meat production. A key barrier in biomedical efforts is that when creating tissues in the medical field, they must be alive and fully functional to work as a transplant. For food, on the other hand, you just need muscle growth. For example, growing a kid-ney that's going to be transplanted into someone's body requires that the researchers get that kidney as near identical to a natural, fully formed and functional kidney, which is a major technological bar-rier. Growing muscle, they noted, just requires taking the cells from skeletal muscles (the kind of meat we typically eat), isolating them, and affixing them to a scaffold that could help anchor them while they proliferate just as they would in an animal's body. Those scaffolds could be made of collagen mesh or even microcarrier beads, all while being rotated in a bioreactor (a fancy word for a steel drum in which cell culture takes place) with electrical stimulation that keeps the cells exercising and warm. The technique they envisioned could produce ground meat only, the authors warned, since the cells in the center of the thicker muscles would be deprived of nutrients and become necrotic in the absence of blood vessels to transport nutrients their way.

And while Matheny's primary goal was to generate interest among tissue engineers, as a grad student at the University of Maryland at the time, he knew his school's public relations department would love the attention this would bring. Their press release did the trick.

"With a single cell, you could theoretically produce the world's annual meat supply," the UMD press release touted. "And you could do it in a way that's better for the environment and human health."

Overnight, Matheny essentially became the face of the cultured-meat movement. Soon, he was being quoted everywhere, from the *Washington Post* and NPR to *CBS Evening News* and *BEEF* magazine, the trade publication for the cattle industry in which he bravely suggested "that perhaps the future farmers of America are microbiologists rather than cattle ranchers."

The *New York Times* profiled him in its annual "Ideas of the Year" feature. *Discover* magazine named in vitro meat one of the most notable tech stories of 2005. When they asked Matheny if people would balk at the thought of eating meat grown in a lab, he rebutted, "There's nothing natural about a chicken that's given growth promoters and raised in a shed with ten thousand others. As consumers become educated, a product like this would gain appeal." Even so, years later, Matheny is still asked the same question in essentially every conversation on the topic: Will anyone actually eat something like this?

The widespread media attention led Matheny to start traveling the country, discussing the benefits of cultured-meat research. He even

managed to get an audience with two of the largest meat producers
on earth: Tyson Foods and Perdue Farms. He suggested they fund their
own R and D and compete against each other to bring the first cultured
poultry to market. Matheny also informed them that the Dutch subsid-
iary of Smithfield Foods—the world's largest pork producer—had sup-
ported cultured-meat research in the Netherlands, and he wondered
whether its counterpart in the United States might do the same.

The poultry producers told him that, while many people think
of their companies as being in the animal production business, they
really see themselves as being in the protein production business. To
them, it didn't matter so much where that protein came from so long
as it was healthy, safe, and nutritious. The thought of getting these
meat Goliaths involved tantalized Matheny. He knew they could
bring R-and-D resources that would dwarf the limited spending by
governments and academia on cultured-meat research so far, so he
made his case and braced himself for the response. The poultry company reps were polite and listened, though at the end of the calls, they
indicated that it was just too early for them to act on this advice.

In many ways, their decision was understandable. This concept
was in its infancy, the science largely theoretical, and the thought that
consumers would even want to buy such meat was far from clear. These
companies already had a proven method of bringing meat to the table
and this method must have seemed far more like something out of a
Jetsons episode to them than a legitimate business idea they'd pursue.

Undeterred, Matheny reached out to other movers and shakers
in the tech and food worlds throughout 2005. He also got a chance
to visit the lab of the NASA-funded researchers in New York that had
sparked his interest in the whole topic three years earlier.

He wasn't sure what to expect, but he certainly anticipated something more grandiose than what he found. The genesis of his interest in growing meat in vitro was hardly a sight to behold. The space
where three years earlier goldfish muscles grew was just two small

tables pressed against each other. Were they dining room tables, Matheny recalls, it'd have been difficult to seat four people.

As he chatted with the researchers, Matheny stared at those two small tables and fantasized about the massive meat breweries he hoped they'd spawn in the near future.

In his meetings with venture capitalists and agribusinesses alike, the biggest pushback Matheny got when describing in vitro meat was that it was just "unnatural." He found this criticism deeply frustrating. "Flying, using email, air-conditioning, reading books, eating foods that grew on the other side of the world—they're all unnatural and extremely new on the timeline of humanity's existence," Matheny points out. "We should celebrate these innovations and appreciate how much better they make our lives."

Still, it's just very difficult to shake that initial reaction to hearing about something like growing meat in a lab. In 2005, the European Commission polled residents about their views on potential future applications of technology, asking if they approved of a variety of applications in some, all, or no cases. Perhaps in response to Matheny, the commission included a question about whether Europeans approved of "growing meat from cell cultures so that we do not have to slaughter farm animals." More than half the respondents said they'd "never" approve of it, though a quarter said they'd approve in some or all cases. Shockingly, more people approved of "developing for children a genetic test that would identify their talents and weaknesses," and even "using genetic testing to produce a child that could act as a bone-marrow donor" than they did producing meat in a lab. It's possible that this skepticism among consumers stemmed from their lack of awareness of the technology—after all, back in 2005, Matheny was one of very few people championing research into the field and no one (besides the participants in Oron Catts's frog legs art exhibit) had tasted meat grown outside an animal. It's also possible that the framing of the question influenced more negative answers since, as we'll see later in the book, more recent polls that provide better context on the issue have been met with far greater support.

Regardless, in light of findings like this, and as Matheny did more and more media interviews, it became clear to him that one reason so many people were grossed out had to do with terminology. While he kept calling the hypothetical food "in vitro meat," which was scientifically accurate, he realized it was akin to calling table salt "sodium chloride"—technically correct, but hardly enticing. Whenever he'd refer to "in vitro meat," people immediately thought of in vitro fertilization—and not that many people want to think about babies while contemplating the meat in their sandwich. Matheny needed a new name for the meat he hoped consumers around the world would one day enjoy.

In the same way he used friends as a focus group for naming New Harvest, Matheny went back to his informal focus groups to brainstorm a better name for the meat. "Lab-grown meat," "test tube meat," and "synthesized meat" all fell into the same category, producing an "ick" factor that immediately prejudices people against the food. One suggestion, to appeal to environmentally minded eaters, was "green meat," though it quickly became clear that it, at best, conjured images of Dr. Seuss and, at worst, rotten meat. As a play on "in vitro meat," one friend jokingly suggested simply calling it "in meatro."

For a time, Matheny favored "hydroponic meat." After all, millions of Americans by that time had become accustomed to buying hydroponic tomatoes, and some even correctly associated them with lower water use. But it was still just too technical. It wasn't hard to envision tomatoes growing without soil, but meat growing without an animal? Even more entertainingly, one friend reminded him at the time that a whole generation of young people, thanks to Snoop Dogg, had a very different connotation of the word "hydroponic."

"Meat without feet," "good meat," "cultivated meat," "clean meat"— the list went on. In an appeal to history, "Churchillian meat" even got discussed, though associating the food with a man who'd been dead for decades wasn't the most popular option. Years later, in a tip of the hat to an episode of *The Colbert Report* that featured the issue and dubbed it "schmeat"—as in "meat schmeat," or sometimes "sheet meat," or, to Colbert, "shit meat"—in 2013 Oxford Dictionaries even awarded "schmeat" as its runner-up new word of the year.

In the end, Matheny's small group of friends settled on "cultured meat." Americans were used to eating cultured products like yogurt, beer, and sauerkraut, and the term brought connotations of good digestive health as well as a sense of refinement to contrast it with the lower-grade conventional meat. "In vitro meat" had earned its place in the history books, but Matheny now felt it was time to put it to rest. (For his part, Willem van Eelen was opposed to any name other than simply "meat," since he argued that's exactly what it was, and it required no special designation.)

Partially as a result of Matheny's conversion to "cultured," for the next decade, essentially everyone in the cellular-agriculture community came to adopt the term. In fact, at a 2011 Swedish conference that Matheny helped organize, the leading researchers in the field officially agreed to the switch. Since then, those in the industry attend symposia with names like the International Conference on Cultured Meat and researchers publish papers with titles such as "Cultured Meat from Stem Cells: Challenges and Prospects." And if you type "in vitro meat" into *Wikipedia*, it mercifully automatically redirects you to its page entitled "Cultured meat."

But six years after the official name change, some of the food's advocates weren't so convinced that "cultured" really was the best term to use. Although far better than "petri dish meat" and "lab-grown burgers," "cultured meat" can be confusing to consumers who think it's cultured like cheese or yogurt, and more important, a lot of people just have a negative reaction to the term "cultured meat."

As the field widened beyond meat and into leather, eggs, milk, silk, and more, "cultured-animal products" occasionally got displaced

by the more interesting and more accurate title of "cellular agriculture." New Harvest hosted the first-ever conference on the topic, in
2016, entitled "Experience Cellular Agriculture," and some began
wondering if that might even become a moniker for the food: cellular
meat, cellular eggs, etc. Ronen Bar of SuperMeat, an Israeli cultured-
meat company, who was at the conference, joked to me at the time,
"Cellular meat? You may as well call it cancer meat."

Critically, though, no one had ever done any actual consumer
testing of the issue. "Cultured" came to be the term of choice because
the scientists working on the issue thought it sounded best, but no
polls or focus groups had ever been conducted. That is, until 2016,
when the Good Food Institute conducted the first consumer poll to
determine what might be the best term to use when talking about
this new technology to the public. The poll tested the five terms offered for the survey by the leading scientists in the field: "cultured
meat," "pure meat," "clean meat," "safe meat," and "Meat 2.0." (No one
even suggested "cellular meat.")

The results were pretty stark. In the two surveys GFI conducted,
"cultured" ranked fourth out of five in terms of consumer acceptance.
In first place was a term Matheny had considered in 2005 but ultimately decided against: "clean meat."

Interestingly enough, it turns out that as far back as 2008, there
were efforts to start calling it "clean meat." Wesleyan psychology professor Scott Plous published a letter to the editor in the *New York
Times* in which he made that case. Bristling that the *Times* had referred to it as "fake meat," Plous protested in his letter: "The commercial development of meat from animal tissue won't result in 'fake
meat' any more than cloning sheep results in fake sheep. Quite the
contrary, lab-based techniques have the potential to yield far purer
meat, uncontaminated with growth hormones, pesticides, *E. coli*
bacteria, or food additives. A more accurate name for the end result
would therefore be 'clean meat.' "

GFI's Bruce Friedrich argued to his colleagues in the field that the term "clean meat" is similar to calling renewable power "clean energy." The general category, clean energy, comprises various kinds of earth-friendly energy sources: solar, wind, geothermal, etc. And since growing animal products requires so many fewer resources and causes so much less climate change than raising and slaughtering animals, the clean energy comparison seemed apt.

More important, asserts Friedrich, the food safety benefits of this meat—namely the lack of intestinal pathogens like *E. coli* and *Salmonella*—make the "clean" label even more appropriate. Unlike conventional meat, which is typically so riddled with bacteria that you have to decontaminate countertops that raw meat has touched, clean meat is perfectly safe to handle in raw form, with a greater risk of contamination coming from your own hands than the meat itself.

Upon beginning to use "clean meat" with the public, Friedrich noticed a much better response than when he used to use "cultured." The "eww" response he'd often get when telling people about culturing meat outside an animal was replaced by people asking what made it cleaner, enabling him to discuss the meat's benefits as opposed to merely its method of production.

I witnessed anecdotally what Friedrich was describing firsthand, at a conference I helped organize in Washington, DC, called "The Future of Food." Friedrich was on a panel with Susie Weintraub, an executive vice president of strategic marketing and business excellence for Compass Group, the largest food service company on earth. In 2016 *Fortune* magazine named Weintraub "one of the most innovative women in food," and she's often regarded as among the most powerful people in the food industry. When Friedrich talked about why GFI favors "clean meat" rather than "cultured meat," Weintraub had an instantly positive reaction. She exclaimed to the crowd, "I'm happy to hear that we've shifted to clean meat versus cultured meat. . . . It's just these little things, like something as simple as going from meat grown in a lab—what the hell, right?—to 'clean meat,' a much better term. People are much more accepting of it."

Quartz published a story in 2016 about Friedrich's crusade to change the name with a title he admits was far from productive: "To Lure People Put Off By the Freakiness of Lab-made Meat, This Is What the Industry Wants to Call It." But the story made a good point. Journalist Chase Purdy noted:

> *Research suggests the biggest influence on a person's opinion of a particular food is how they "expect" it will taste. Giving fun, enticing names to healthy foods increases the desire to try them. Why not call broccoli "broccoli bites" or carrots "X-ray vision carrots"? Renaming foods to make them sound more appealing resulted in an increase in the sale of vegetables in the school cafeteria by 27 percent.*

Subsequent polls and focus groups conducted in 2016 by Animal Charity Evaluators and in 2017 by New Harvest both confirmed what GFI had found: "clean" substantially outperformed "cultured," leading most of the companies in the cell-ag field to switch from "cultured" to "clean."

Naming debate aside, there are still many things that need to happen before clean meat becomes a viable consumer product—let alone popular enough to transform our food industry. New Harvest's primary efforts in its early years involved helping organize European conferences and other events on cultured meat in an effort to increase awareness and attract sources of funding. But with Matheny running the organization alone in his spare time while he went to school and worked, no monumental progress ensued. Not a gram of meat was produced, no companies were formed, and the dream of getting meat on store shelves still seemed distant. After graduating in 2009, now having obtained numerous academic degrees—BA, MBA, MPH, and PhD—Matheny began working for the Intelligence Advanced Research Projects Activity (IARPA), a federal organization.

Convinced that technology can vastly improve welfare, and that the only real threat to technological advancement is a global catastrophe, Matheny focused more of his energies on his work at IARPA to reduce risks from war, pandemics, and technological accidents. At the same time, as he felt like he was neglecting New Harvest, a molecular and cell biology student in Canada, Isha Datar, had written a paper on the potential for cultured meat and sent it to Matheny for his thoughts.

In 2010, the journal *Innovative Food Science and Emerging Technologies* published Datar's article "Possibilities for an In-vitro Meat Production System." The IVM name just wouldn't go away, but Matheny was still thrilled to see more serious academic interest in the topic. Because of her zeal, Datar quickly began representing New Harvest in venues around the globe. In 2012, Matheny appointed Datar the executive director—and first-ever employee—of New Harvest. After gaining attention as a 2013 TEDxToronto speaker, Datar attracted enough resources to the organization that New Harvest started giving out grants to researchers and putting on its own conferences.

As we'll see in later chapters, two of the companies in this book, Perfect Day (which makes milk) and Clara Foods (which makes egg whites), were cofounded by Datar, and some of the research now being done to solve key barriers to commercialization is being funded by New Harvest.

"The obstacles to disrupting animal agriculture with cellular agriculture aren't lack of expertise and certainly not lack of interest," Datar notes, sitting in New Harvest's modest New York City office. "The biggest deficiency is simply a lack of funding. Nearly all the funding for tissue engineering research is going into medicine, not food. We need to change that."

Toward that end, Datar created the New Harvest Cultured Tissue Fellowship, a collaboration with Tufts University in which one student will study in the school's Tissue Engineering Research Center as a postgrad. At the end of her studies, Natalie Rubio, the first fellow in the program, will hold the first-ever PhD in cellular agriculture.

When pondering whether people will eat the meat she's attempting to bring into the world, Datar feels confident. "If we're comfortable treating farm animals like bioreactors, and selectively breeding them for the purpose of maximal muscle growth, why wouldn't we just remove the animal altogether and just go for the muscle growth on its own?"

At the same time, Datar points out that the cellular agriculture revolution she's aiming to assist is about so much more than just food. Already there are companies making cultured leather, spider silk, and even musk perfume—all without the animals, and these products could be just the introduction to clean-animal products the public needs to get used to the idea. In so many ways, just like transportation and home lighting, industries that have relied for centuries on animal use now face an oncoming wave of start-ups seeking to make the current models obsolete.

Matheny is now IARPA's Director but still sits on the board of New Harvest. For his part, he self-reflects while sitting in a suburban Maryland burrito shop in 2017. Looking down at his $6 rice-and-beans feast and the meat-filled burritos of other diners around him, he wonders how long it'll take before their burritos will be filled with clean meat.

"We can use technology to render some of our most pressing difficulties moot," he argues. "The habit of high meat consumption is a serious problem that many people just have a hard time breaking. But the cultured meat industry now has a chance of being able to provide people with the same food—probably even better food—without causing so many problems. If I played a small role in helping that happen, little would make me happier."

3

(GOOGLE) SEARCHING FOR A SOLUTION

Through their work at New Harvest, Jason Matheny and Isha Datar certainly have played major roles in helping bring the clean-meat concept into the limelight. But despite their success at increasing awareness and funding for research, they're not interested in actually making any meat themselves. In other words, New Harvest is more like a foundation touting the potential benefits of the internal combustion energy, not a Henry Ford who developed a way to bring this technology to the masses.

Datar, in her tenure at the helm of New Harvest, has even argued that the technology to culture meat (though not egg whites and milk, as we'll see in chapter 7) is still so new, that dollars today are better spent funding academic research as opposed to investing in entrepreneurs. "Start-ups keep their intellectual property private," Datar says. "It'd be a shame for any one company to control the IP on how to grow meat. In my opinion, at this point in time, open-source academic research will do a lot to advance the science of cultured meat. Once the base technology is advanced enough, then we can get into competition." Because of this view, New Harvest now bills itself as a research institute, and its staff of three acts as support and a funding source for its team of research fellows.

This argument wasn't lost on Sergey Brin, the cofounder of Google as well as the Brin Wojcicki Foundation, a nonprofit he started with his then-wife, Anne Wojcicki, to fund projects they consider good for the world. From advancing driverless cars to supporting research into asteroid mining, Brin doesn't shy away from bold efforts. Combine that with his longstanding concern about environmental degradation, and it's clear just why reforming the meat industry would be on Brin's bucket list.

"There are basically three things that can happen going forward," Brin predicts. "One is that we all become vegetarian. I don't think that's really likely. The second is we ignore the issues and that leads to continued environmental harm, and the third option is we do something new."

There's no shortage of such "new" recommendations from sustainable food advocates for fixing our broken protein production sector. Going local, grass-fed, organic, shopping at farmers' markets, and more, all have been put forth as possible solutions to the ills of factory farming. And while efforts on the part of animal welfare, environmental, and food advocates have met with some success, conventional meat production (i.e., factory farming) still dominates animal agriculture in America. Nearly all animal products produced in the United States continue to come from concentrated animal feeding operations (CAFOs), primarily because most people simply buy what's cheapest. There may be a niche segment of the population that heavily factors ethics or the environment into its food choices, but when it comes down to it, most of us make the majority of our food choices based primarily on price, taste, and convenience. Yes, there's increasing knowledge about the planetary harm involved in industrial meat production, but that knowledge so far has done little to substantially dampen demand for CAFO-produced meat.

For those who regularly shop at their local farmers' market or pay a premium to buy organic, locally sourced products, it's useful to remember that Walmart sells a full 25 percent of groceries in America, and a 2013 Gallup poll found that eight in ten Americans eat fast food at least monthly, with half of us eating it at least weekly. This is despite the fact that the same poll found that more than three-quarters of Americans say the food served in fast-food restaurants is "not too good" or "not good at all for you." And lest you think the lines in fast-food restaurants are comprised of those who can't afford anything better, the Gallup poll actually found that the lowest income respondents (earning less than $20,000 a year) ate the least amount of fast food, while higher earners (more than $75,000 a year) ate the most.

The point of mentioning this isn't to knock farmers' markets or other sustainable foods efforts. Rather, it's merely to observe that trends like locavorism, organic, or GMO-free may get a lot of headlines, but that attention hasn't translated into major purchasing shifts among a large portion of the population. The percent of meat sold in America at farmers' markets is barely perceptible, and consequently so is the portion of farm animals in America raised solely on pasture. In other words, most Americans are quite content simply to buy cheap meat from fast-food companies and big-box grocers much of the time.

Some natural foods enthusiasts will balk at a high-tech solution like cellular agriculture, and clean meat may end up not being for those folks. If they view it as too sci-fi to fit their definition of "natural," they'll still have plenty of options, just as they do today. But given that mainstream meat consumers (i.e., most people in our society) rarely think twice about buying meat from animals who've been raised in unsanitary and inhumane conditions, produce that's been sprayed with synthetic pesticides, or processed foods made with genetically engineered corn, it's difficult to imagine most of us having a problem with clean meat that (aside from being safer, eco-friendlier, and more humane) is pretty much the same meat that we're used to eating.

Another advantage of clean meat is, of course, the fact that it doesn't require an animal to suffer or be killed to produce it, something that surely would appeal to at least part of the natural-foods crowd. Even many people who eat conventional meat express concern over how that meat gets to their plate, especially the details regarding just how that animal lived and died. This, in fact, is one of the reasons that Brin was attracted to culturing meat in the first place. "When you see how these cows are treated, it's certainly something I'm not comfortable with," he says.

Realizing the potential benefits that clean meat could bring—not to mention the commercial promise of a technology that could offer a viable alternative to industrialized meat production—Brin set out to find promising scientists in the field whose research he could fund. As a result, in 2009, his foundation contacted Matheny. It turned out that Brin had seen news coverage of the Norwegian conference the year before that Matheny helped organize. Against his wishes, it was called the In Vitro Meat Consortium (even a *New York Times* columnist at the time joked, "They might want to do something about that name").

Matheny offered on the spot a list of scientists now working in the space, but one stood apart from the list: Dr. Mark Post of the Netherlands, a man who was already known for his research by the small group of other scientists in the field, but who in a few years would become far better known than Matheny as the new face of the cultured meat movement.

A physician in Holland who specializes in making tissues in vitro, Post had been thinking about doing the same thing with cattle muscle cells for some time. Unlike Matheny, Post is hardly a vegetarian: his typical lunch consists of a ham-and-cheese sandwich, which he enjoys at his desk inside a medical school building with a gigantic neural synapse adorning its exterior. However, he does share Matheny's concerns about the lack of sustainability in today's meat industry. "I'm really a single-issue voter," the avuncular professor notes in his modest Dutch office. "I vote for whoever will be better for the environment. After all, there are good and bad economic times, but without our planet, what have we got?"

Post's foray into clean meat came several years prior to the contact from Brin, when the Dutch government became the first in the world to start funding research into the topic. Already interested in the issue, Post was thrilled to be part of that first research group.

Working at Maastricht University, a small Dutch college on the Belgian border, Post joined a motley crew of researchers who shared his interest. Hailing from Amsterdam, Eindhoven, and Utrecht, his colleagues tended to devote only one day a week to the project, since they were all involved in other pursuits, but Post quickly got hooked. The scientists involved had different motivations—some were interested in increasing animal productivity, for example—but Post was bent on culturing animal muscle to solve what he sees as the global food crisis.

The only other person involved in the project at that time who shared Post's passion was food chemist Peter Verstrate. A veteran of the meat-processing world, Verstrate was introduced to the concept of culturing meat in March 2003 when, then an R-and-D executive in the European office of Sara Lee Foods, he had an unscheduled visit from Willem van Eelen, the Dutch scientist who had previously caught Jason Matheny's attention for his efforts to grow animal muscle tissue in vitro. Van Eelen had just secured his patent (the first of its kind) and was now looking for money to set up a pilot plant to start making meat.

"The potential advantages of growing meat this way are evident,"

Verstrate wrote in his notes about the meeting afterward. "Animal welfare is no longer an issue; cost can come down; efficiency (of the transition of nutrients in meat) is better; environmental advantages."

The idea so intrigued Verstrate that he authored a memo to both his CEO and managing director arguing why Sara Lee needed to get on board with this idea. The memo began by acknowledging the difficulties. "Getting there will take years and also large investments in people and financial resources." But, Verstrate explained, if they really believe that this is a viable method of meat production, it'll certainly become commercialized and the company should get in at the very beginning, rather than allow its competitors to take the advantage. "That way we stay in control, so we can, when raw material becomes available, be in some sort of preferred position."

After explaining the benefits to his superiors, Verstrate's memo delved deeper into the potential implications of this technology to advocate that Sara Lee get financially involved: "The societal support for starting a development that has the aim to make an end to systematically slaughtering animals will be huge. I can think of many influential entities that would back this."

At first, Verstrate's communication fell on deaf ears. Sara Lee's top brass just wasn't interested in an idea that seemed more like a Mary Shelley novel than something a meat company seeking actual profit ought to be doing. But the food chemist persisted, wearing down his superiors' skepticism, or at least their resistance, until Sara Lee agreed to become a corporate partner in a Dutch government-funded experiment—the same one that Jason Matheny had successfully lobbied to fund.

The research Van Eelen was pursuing never resulted in any real breakthroughs, and the pioneer went to his grave in 2015—at ninety-one years old—without seeing his dream of slaughter-free meat on grocery shelves realized. Numerous obituaries written of his life called him the "Godfather of In Vitro Meat," and one included a photo of Van

Eelen tipping his hat to a cow. Toward the end of his life, he reflected to *The New Yorker*, "I like meat—I never became a vegetarian. But it is hard to justify the way animals are treated on this planet. Growing meat without inflicting pain seemed a natural solution."

Even though they had the same goals, the approach the Dutch researchers would use was quite different from the unsuccessful efforts of Van Eelen. Rather than starting with an actual visible piece of tissue and trying to coax it to grow, as Van Eelen did, Post and his colleagues started at the cellular level.

The physician focused on one type of cell in particular, myosatellite cells, which are the precursors to skeletal muscle cells—the kind composing the meat we typically eat. Unlike certain other kinds of cells, these myosatellite cells have only one potential career path: to become muscles. In other words, unlike stem cells, which can metamorphose into any kind of cell in the body, myosatellite cells simply wait in your body until there's a need to grow muscle—say after a tough workout in which muscle fibers have been broken down, for example—and then go to work performing their singular task of growing more muscle.

And those muscles, Post predicted, would be the key differentiating factor between his meat and the numerous plant-based meats already available that maintain only a tiny fraction of market share compared to animal meat. Post believes that the plant-based meats "are usually more expensive than meat and don't yet perfectly mimic the taste, even though some are pretty good. We're committed to producing a natural meat product that is both cheaper than farmed meat, indistinguishable in texture and taste."

In 2009, just one decade after Van Eelen filed the first-ever patent on a cultured meat process, Post prevailed in growing mouse muscles in vitro. He'd done just what he envisioned himself soon doing with beef: he'd isolated myosatellite cells from the rodents' muscles, anchored them in a petri dish and allowed them to flex on their own, growing bigger and stronger muscle fibers along the way. "You essentially give instructions to cells by starving them," Post explains. "That causes them to grow, and the anchors allow the cells to contract, causing tension, which causes protein synthesis."

The research was part of the same study funded with $2 million euros by the Dutch government, and even though the little mouse muscles weren't impressively sized—just twenty-two millimeters long, eight millimeters wide, and half a millimeter thick—their importance to Post was massive. And the success piqued Brin's interest.

"From the beginning, I was fascinated by the concept of creating meat in a laboratory, and it turns out so was Sergey Brin," Post delights.

Having reviewed Post's success with the mouse cells and read a media interview, Brin had Rob Fetherstonhaugh, a Canadian representative from his foundation, ring the tissue scientist up to inquire about his and Verstrate's pilot project. Fetherstonhaugh didn't identify himself as an associate of Brin; for all Post knew, he was just someone from Montreal who shared his optimism about culturing meat. But when Fetherstonhaugh asked if he could meet Post in person in the Netherlands, the physician suspected the caller might have more than just a passing curiosity in the subject.

And so it was that on May 5, 2011—a national holiday in Holland—the two men sat down in Maastricht and Fetherstonhaugh revealed to Post the name of his boss. "At first he just kept calling him Sergey," Post laughs later in his office. "I could tell by the way he was saying it that I was supposed to know who this was, but I finally had to admit that I didn't know this Sergey."

Once it became clear just who was really asking about his work, Post's mind started racing with the possibilities. "It was so clear to me that we could do this. The science was there. All we needed was funding to actually prove it, and now here was a chance to get what was needed."

Fetherstonhaugh requested a funding proposal from Post for a proof of concept, two pages maximum, and asked for it within a week. The scientist smiled, assured his visitor that would be no problem, and got to work. As soon as Fetherstonhaugh walked out, Post called Verstrate elated and eager to share the news.

"I knew it was going to succeed," Post recalls. "I never expected it could fail. We had to go through the motions and actually produce the meat, but this wasn't experimental to me. This was the real deal that was necessary to get us to prime time: commercialization."

After the Dutch government–funded study ended, the researchers brainstormed what they thought would be a huge hit with the press. "Our original idea was to take a biopsy from a pig on the Maastricht campus, grow a sausage, and then hold a press conference where people would eat the sausage while the same pig—still alive!—would be walking around on the stage. Wouldn't that have been great?"

Post's idea for an in vitro sausage, while perhaps most appealing to a Dutch audience, didn't tickle Brin's fancy as much as an all-American hamburger. "Even though he's from Russia, he knew that the hamburger is king in America, and so we had to go with a cow."

The proposal estimated it would cost about $330,000 to produce the world's first cultured burger. But to paraphrase one of the characters from the movie *Contact*, why build one when you can build two for twice the price? Brin offered nearly three-quarters of a million dollars to the team to make two cultured burgers, and the work began.

The basic process they planned to use requires, generally speaking, four-steps: (1) extract the myosatellite cells from the cow via a simple biopsy; (2) place them in a nutrient-rich medium that will feed the cells and allow them to divide and grow in number; (3) exercise the cells with electric currents so they become actual muscles and keep gaining mass; and finally (4) harvest the meat and perform any further processing needed, such as adding fat or other flavors.

It should be noted that, even though the cells Post and Verstrate used to grow the meat only required a small sample of cells from a cow, the process of culturing those cells at that time wasn't slaughter-free. Since humanity began culturing cells in labs 130 years ago, the medium being used to provide muscle cells with the nutrients they need in vitro has typically come from blood. In Post's case, the medium is known as fetal bovine serum, and the grisly process used to obtain it involves slaughtering a pregnant cow and removing the fetal calf from her carcass. The serum is then extracted directly from the fetal blood and used in the lab to keep the dividing muscle cells alive in vitro. Depending on how much serum is used, some estimates find that one cow fetus can provide enough serum for just a single kilogram of meat, begging the question of where all these calf fetuses would come from if cultured meat were to be commercialized while still requiring such serum.

It's not exactly the kind of thing that comes to mind when you think "cruelty-free" or "sustainable." Nor is it cheap. Just one liter of fetal bovine serum can cost around $500, making it a financial, as well as an ethical, problem. Fortunately, several cellular ag companies have already completely done away with such serum, typically by using plant-based or synthetic serums or by simply figuring out ways of going serum-free, as have others. (Interestingly, the NASA-funded researchers who published in 2002 found comparable growth when they used a maitake mushroom extract rather than fetal bovine serum.) But back in 2013 at least, Post was still using fetal bovine serum, since he simply wanted to provide a proof of concept that you could actually grow enough meat outside an animal to make a meal.

Post figured he needed to grow about twenty thousand bovine muscle fibers from his starter cells to have enough meat for a burger.

At the rate he was going, that would take only three months—much faster than a cow would take to reach slaughter weight. And had Post had more staff and space, the process could've taken mere weeks. Most beef cattle are slaughtered at around fourteen months old, though grass-fed cattle (those who never live in a feedlot) typically take about twenty-four months to reach market weight. In other words, no matter how you slice it, Post and Verstrate can grow cattle muscle a whole lot faster than a cow can, and once at scale, a whole lot more muscle than any herd of cows could in that same time. And given that they're only growing muscle and not the other parts of a cow that we want less, it requires far fewer resources to produce. On a feedlot, for example, one cow will eat more than twenty pounds of feed per day. Those days add up, and make it clear why so much of the world's cropland is devoted to growing corn or soy that will be turned into feed for livestock. In other words, if you want to know why the Amazon is being chopped down, you needn't look much further than the world's growing demand for meat.

Maastricht University explained in a statement why it was sponsoring the futuristic project: "Humans as a race have shown no sign of wishing to eat less meat, so it is unrealistic to think about eradicating meat from the human diet in the future. A sustainable way of providing it has to be found."

Armed with his university's support and Sergey Brin's funding, Post was ready to start growing some beef. Followed by a film crew, Post went to a small slaughterhouse about three miles from his office in order to obtain the cells he needed to start the culturing process. The facility, which Post describes as "boutique as slaughterhouses go," only accepts nearby pasture-raised cattle who were never fed hormones or antibiotics. And unlike typical slaughter plants, which kill up to four hundred cattle per hour, the rate here is one animal every ninety minutes, since the sole two employees fully butcher each animal prior to bringing another cow inside.

The slaughterhouse owner had agreed to let the scientist take a miniscule muscle extract from a Belgian Blue—a breed known for its lean and muscular physique, which makes them look much more Arnold than Bessie. Unfortunately for Post, taking a biopsy from a live animal for lab purposes is considered an animal experiment in the European Union, so in order to avoid delays with the project—if not thwarting it altogether—Post had to take his sample from the cow after she was slaughtered.

"It was a very small spoonful amount of muscle that we needed." Post demonstrates by placing his finger and thumb less than an inch apart. "I wish we could've used a biopsy from a live animal, though," he regrets, "but there's absolutely no difference between the muscle of the live animal versus a just-killed animal. You could easily take a biopsy from a living cow and do exactly what we did if you had the permission to do the biopsy."

Using this culturing process, Post calculated, one sample from one cow could, in theory, produce twenty thousand tons of meat, or more than four hundred thousand cows' worth of beef—enough to make about 175 million Quarter Pounders. In a way, this sample could be thought of as cellular Eves creating generations of new cells that could save the world one burger at a time.

But first, Post had to make one.

It takes just over two hours to go from meat biopsy to a started cell culture. Then it takes about thirty hours for the first stem cell to divide into two. And from there, another thirty hours for two to become four. Then thirty hours later you've got eight, then sixteen, thirty-two, and very soon: millions. The microscopic stem cells Post and his team collected reproduced quickly, and importantly, they created more muscle fibers in the process. As noted, these myosatellite cells are the same kind of cells that repair a cow's (and our own) muscles after an injury, making them a seemingly perfect fit for the task at hand.

Having already succeeded with the mouse muscles, he never doubted for a second that he could do the same with cow cells. But having conviction is one thing; seeing his cells proliferating in real life would be another matter altogether. The suspense ate at Post while the cell line was getting established. Even with his earlier success, the question of "what if" something went wrong or unexpected hid in the back of his mind. Patience would have to be his virtue.

When the cells began organizing around the anchor in the center of the petri dish, giving them shape, Post could hardly contain his excitement. Feeding them their steady diet of amino acids, fats, and sugar was like feeding his own baby. Post imagined his little cells happily gorging themselves on the nutrients he supplied. In the beginning, there was just a mere blob to organize, but within two days, muscle was already visible between the anchors. Post's paternal pride grew just as quickly as the cells.

"And they do that all by themselves," Post marvels, confessing that he'd sometimes just watch the petri dishes for minutes on end, as if he'd be able to see the cellular division occurring with his own eyes.

The mini muscles worked away, building tension and contracting, just like they would in a cow's body. This process, mimicking exercise, creates strength and the muscle begets more muscle. In just a few weeks, the first of that muscle was ready for harvest, later to be combined with the muscle growing in nearby petri dishes. A few thousand more strands that could be layered on top of one another, Post fantasized, and the world would have its first-ever cultured burger. With much trial and error, the team tried to figure out how to turn their muscle strands into some conglomeration that could actually be considered a hamburger.

In early 2013, just a couple of months after he'd started the culture, Post had enough meat now produced and assembled that he knew it was time to privately sample. With no more than a few grams of cultured tissue, he, along with his comrades, now had to play the role of the chef.

The raw muscle started out yellow in its uncooked form, and began to sizzle as soon as it hit the sunflower oil, already hot from coating the small skillet. Soon the aroma of cooking beef filled the air and the tasters started to salivate. Not one nose would be able to detect that there was anything abnormal about this beef. They knew how much was riding on whether this meat would pass an objective taster's test, but first it had to pass their own.

"The main thing I wanted to know," Verstrate recalls, "is if it would cook in the skillet just like conventional meat does. It was the food chemist in me that was more in suspense than the diner. I was so relieved."

They flipped their mini burger from one side to the next, being careful not to burn either side of the valuable patty. To their relief, as it cooked, it browned just like conventional beef.

They removed the burger from the pan, placing it on a plate next to them for it to cool. The moment of truth awaited them. Years of work, first on rodent muscles and now on cattle muscles, had come down to this very moment, and each of them could barely contain their excitement.

"Here we go." Post smiled to his colleagues as they prepared to go where no eater had gone before.

Post, Verstrate, and Fetherstonhaugh divided the sample into tiny pieces among themselves. None of the three thought there was any risk in trying it, but if they were going to get sick from eating the world's first cultured beef, at least they were only eating a miniscule portion.

Choosing not to season the meat so as not to compromise the flavor, the three simultaneously placed their portions on their tongues. They closed their eyes and began chewing—slowly and deliberately.

Each wanted to savor the moment as they became the first humans ever to eat beef grown outside a cow.

After a few moments of chewing and finally swallowing, they all felt fine. It was difficult to taste much from the small volume consumed, but each taster did sense a meaty flavor. Verstrate declared the taste test to be a success, and keeping their eyes on the prize, the team went back to the petri dishes.

A few weeks before the official taste test, in August 2013, Fetherstonhaugh informed Post that Brin wanted to finally meet him in person to chat about how the press conference would go. When Sergey Brin wants to meet with you, you go to him, so Post made the journey to Northern California to at long last meet his billionaire backer. Expecting to travel to Google's storied Mountain View headquarters, Post instead found that Brin was holding office at his child's nearby daycare. The Google cofounder was wearing a T-shirt, Bermuda shorts, Crocs, and of course a pair of Google Glass. What shocked Post the most was the apparent total lack of security surrounding one of the world's richest men. "It was just him and me."

Over a vegan breakfast offered by Brin, Post expressed his gratitude for the investment and went over the logistics for the press conference, and Brin let Post know how glad he was to have helped make this possible. After fifteen minutes the meeting was up, and a communications CEO in a full suit was next in line to get his quarter hour with the tech giant.

Back in the Netherlands, three months after starting the bovine cell culture, Post had enough muscle meat to make his burgers. Removing the small amounts of grown tissue from each of his hundreds of petri dishes, he and the team began painstakingly assembling the various thousands of strands of muscle and placing them into just two petri dishes. Because the meat lacked myoglobin—a protein that helps make mammalian muscle red—it had the relatively colorless look of chicken flesh, as opposed to the deep red we see in beef. To correct this cosmetic problem, Verstrate added a touch of saffron and beet juice to the tissue—enough to tinge them red without impacting the taste.

They now looked like two typical beef patties—albeit the most expensive beef patties in history—and the team was ready to show them off.

They chose London as the location for their press conference, seeking to maximize the number of international journalists who'd be able to attend. As preparations began on just how the tasting would work logistically, the Dutch researchers also deliberated about how to transport their $330,000 burgers to the venue. Given how valuable their cargo was, there certainly wasn't any shipping service they could trust to get the sacred patties from Maastricht to London. Just like protective parents with a child not yet ready to travel alone, Post insisted on being the only person to handle the burgers, even though it was sure to make travel more complicated.

The evening before he traveled to London, Post did what he did at the end of each weekday: he biked twenty minutes to get from his office at the university to his home. But this time, on his bike was perched a cardboard box with cargo valued at nearly three quarters of a million dollars. "Why pollute?" he said when asked why he didn't just take a cab. "My bike works just fine."

It was the hottest day of the year—91 degrees Fahrenheit. Post wasn't sure if the patties could stand such heat, even for a short time, but it wasn't worth the risk. He stored them on ice inside a Styrofoam container stashed inside the cardboard box. Followed by the same film crew that had accompanied him just a few months earlier to obtain the cells that had produced these two burgers, Post smiled as he rode his bike through Maastricht's historic streets. Little did pedestrians gawking at the spectacle know that this man was transporting both a small fortune, and a package that would soon be the center of a global media storm.

After making it safely to his house, when he walked in the door, Post put the box—ominously marked "Not for Human Use"—in his family's refrigerator. The label wasn't meant to dissuade his family from looking inside; it was simply a legal requirement for transporting the material outside the lab. As his head hit his pillow that evening, Post couldn't stop thinking about the box sitting in his kitchen at that very moment. Next to a carton of orange juice and above a drawer of greens sat what was almost certain to become the most important project of his life. Thankfully his kids didn't raid the fridge that night.

On his way to London the next morning, Post first took the train to Brussels, stashing the box in the luggage rack above but in front of his seat. "I wasn't taking any chances," he later said. "Either the box was touching me at all times or I could at least see it up there." Needless to say, he didn't allow himself to sleep on the train.

At Brussels, he and Verstrate made the rest of the journey to London by train together, each taking turns watching the precious cargo.

"We would've just flown"—Verstrate grins—"but we were too concerned about airport security." The two weren't entirely sure if bringing these patties into the UK constituted smuggling or not. They did have documents to bring animal material into Britain, but not for human use, which these very clearly were. In fact, they might soon become the most publicized act of human consumption of hamburgers ever.

When they arrived at customs, Post and Verstrate held their breath. As the customs agent examined their box, they felt secure in their ability to get it into the country. And then the world stopped. The agent's face wrinkled as he looked at the box's "Not for Human Use" warning. "That's odd," he quietly said to himself as he more closely inspected the package.

Post's heart halted; the professor stopped himself from biting his lip and showing any concern. "I'm a scientist at Maastricht University," Post offered, getting ready to pull out his university ID as proof.

The agent waved his hand to indicate further explanation wasn't necessary, perhaps just wanting to move the line along. To their great relief, the agent didn't open the box, and on they went to their hotel. "I almost had a heart attack!" Post jokes in hindsight.

The next decision was how to store the burgers. His own refrigerator had sufficed the night before, and now they had to make it one more night. "We thought about just sticking the patties in the hotel room minibar, but then someone suggested we special order an empty refrigerator to be waiting in the room, which thankfully we did," Post remarks. "It would've been a hassle taking all those little vodka bottles out!"

With the logistics now choreographed, the burgers and their travel-weary custodians arrived safely in West London. As they went over how the event would go down, the media and guest lists were finalized. VIPs from all over Europe and America were flying in for a ringside seat, including about a hundred journalists. One group was conspicuously absent from the festivities: no meat industry executives sat in the audience. The organizers don't recall if any such executives were invited or not, but their absence was certainly noticed. Instead, the very people whose industry stood to be most disrupted by this new technology would have to learn about the event just like the rest of the world: from news reports later in the day. At this point, the idea likely seemed too theoretical and futuristic to meat sector leaders to warrant attendance.

Post created a scene at Riverside Studios that appeared less like a traditional press conference and more like a cooking show, replete with kitchen counter, single-burner stove, and a display sink attached to no plumbing whatsoever.

Not taking any chances, Verstrate and Post allowed the chef to cook one of the patties before any spectators entered the room in order to get a feel for how it would fry up. "We weren't risking this guy burning our big opportunity, literally and metaphorically!" Verstrate says. To their relief, the burger fried up perfectly and was placed off to the side. That patty—at first far less famous than its successor—has now been plasticized and placed in the Netherlands' Museum Boerhaave, a museum specializing in the history of science, aptly next to the world's first microscope, also invented by two Dutchmen.

"I do dream that one day it'll be moved to a museum that chronicles the rise and fall of the animal-agriculture industry," Post says.

Media trucks with large satellites perched on their hoods idled outside the studio. Journalists clad with "Cultured Beef—Press Pass" badges interviewed onlookers in the lobby.

In a moment of candid transparency, Post turned to the film crew documenting the spectacle and let out his fear. "The scary part," he opened up, was that if the tasters didn't like the burgers, "I would look foolish, and I would be responsible for pushing back this field." It wasn't all scary, though. Reflecting on how many photos were being taken of his hand holding the burger, Post later joked to me how glad he was that he'd recently cut his nails and hadn't worked on his bike in several days.

When the time came, the doors to the studio opened, and like commuters bustling by one another to get on the subway, journalists jockeyed for the best seats in the house. After the audience was in, Post surveyed the room from offstage. There wasn't an empty seat in sight. His dream was coming true. Taking a deep breath, he grinned and walked onto the makeshift kitchen set.

"What we are trying today is important because I hope it will show cultured beef has the answers to major problems that the world faces," Post announced. "Our burger is made from muscle cells taken from a cow. We haven't altered them in any way. For it to succeed it has to look, feel, and hopefully taste like the real thing."

Before him sat a silver dome atop a porcelain plate, and to the side, a frying pan and a plate ready with sesame seed bun, romaine, and sliced tomatoes. Those sides were merely props, though—these tasters weren't about to obscure their palates with anything that could mask the taste of the most expensive meat ever produced.

And so stepped in famed chef Richard McGeown, ready to extract a sole beef patty from its petri dish home and start the show. Adding sunflower oil to the already-hot pan, he placed the products of years of hopes and labor into the heat. It sizzled just like a conventional burger would, and pretty soon, just as it had during the private test run, the aroma of cooking meat filled the air.

Once the patty was browned on each side, the two testers Post had selected to perform the taste test were invited to try it. In deciding who to select for this opportunity, Post concluded he wanted people who were outside the movement enough so as not to appear biased but who were also considered authorities on food. The first was an American food writer, Josh Schonwald, author of *The Taste of Tomorrow*, a book about the future of food. In doing his research for the book, Schonwald had visited Post in 2009 to talk about cultured meat, though at that point there was no actual meat to sample. The second tester was Hanni Rützler, an Austrian lecturer and author who specializes in future food theorizing and had published numerous articles and books with predictions about how future humans will sustain themselves.

Rützler listened to Post extoll the virtues of this burger to the audience, patiently waiting her turn to be in the spotlight. One problem, however: Post wouldn't stop talking. As he enthusiastically described the process and promise of the burger, it was quickly losing heat, and who wants to eat a cold burger? She decided not to wait any longer.

Holding the burger down with a fork in her left hand, Rützler cut about a fifth of the patty with the knife in her right. The feel of the cut was just like that of a conventional burger, she later remarked. Now with her prize skewered onto her fork, Rützler raised the meat to her nose and smelled it. Post continued touting the benefits of cultured meat, not even looking at his taster as she prepared to make history.

Again she smelled it, now visually examining it with a curious look after her second whiff. And as Post continued evangelizing, sitting in front of a bank of cameras, Rützler put the culmination of his culturing career into her mouth.

She closed her eyes and took twenty-seven bites. No one, it seemed, was still listening to Post. All attention in the audience was trained on the taster, even as Post continued talking, still unaware that she was already chewing.

Finally, her verdict.

"There is quite some flavor with the browning," she said after taking a moment to consider the experience. "I know there's no fat in it, so I didn't really know how juicy it would be, but there is quite some intense taste. It's close to meat; it's not that juicy, but the consistency is perfect."

After his own taste test, Schonwald declared that the mouthfeel and texture was pretty similar to that of a conventional burger. It was "kind of an unnatural experience," he deadpanned about the lack of ketchup, but his final judgment: it brought to mind less of a burger experience and more of something akin to what he called "an animal protein cake."

Talking with the BBC later in the day, Verstrate noted that the product was still in the early stages of development and not yet ready for mass commercialization. "It consisted of protein, which is muscle fiber. But meat is much more than that—it's blood, it's fat, it's connective tissue, all of which adds to the taste and texture." In other words, what the tasters had tested wasn't actually identical to a hamburger from a slaughtered cow. This prototype burger was just pure muscle, whereas once such burgers are commercialized, they'll have fat added to give them the same mouthfeel as conventional meat.

Thinking aloud all the possibilities, Post forecasted to a small group of people how this could revolutionize our lives. "Theoretically, we could even create hybrids of animal muscle cells. Say you wanted a lamb-tuna steak, we could probably mix their muscle cells together." Kind of like taking the turducken to a whole new level—the cellular level.

At the end of the press conference, as attendees began chatting with one another, several of the journalists noticed that half the burger was still left uneaten in the fake kitchen setup. A vast amount of money—not to mention symbolic importance—was just sitting unattended on the counter. Some begged and cajoled Post to let them sample it themselves, but the doctor rejected them all, smiling and informing them that he'd already promised the leftovers to his kids.

"Today we've seen proof for the first time in the world that you can actually make meat from cells taken out of the cow but not produced inside of the cow," Post proudly declared right after the tasting. "It's important also to have gotten the message out that we really have to come up with an ethical and environmentally friendly way to produce meat."

Even with the measured tasting reviews, the press conference was a hit. The *Washington Post* headlined, "Could a Test-tube Burger Save the Planet?" *The Economist* entitled its story, "A Quarter-Million Pounder and Fries."

The tasting ended up winning Post accolades and awards worldwide, including the World Technology Award, a prestigious honor bestowed upon those engaged in groundbreaking work of "the greatest likely long-term significance."

All the media coverage even spawned the first-ever *In Vitro Meat Cookbook*, perhaps the only cookbook ever sold exclusively with recipes that no user today can make. Published in 2014, the cookbook has recipes for all types of newly imagined foods that cellular agriculture might make possible, from knitted meat to "celebrity cubes" of flesh from, yes, your favorite celebrities. As the cookbook authors quip in their marketing of the cubes, "Forget autographs or posters. Prove that you're the ultimate fan of a celebrity by eating him or her. . . . If you can't be famous, at least eat famous." And of course, for those who were intrigued by the scene in *Hannibal* where Dr. Lecter pan-sears and then serves Ray Liotta's character pieces of his own brain, the *In Vitro Meat Cookbook* offers a recipe for "In Vitro Me." (Why settle for anything less than yourself?)

Calls came in from far and wide. Investors and scientists alike wanted to know how they could get in on the action. A common question Post heard was whether this was just a novelty item akin to the (literally) astronomically priced future commercial space trips now being sold, or something that could actually be scaled up to compete with the conventional meat industry.

The science is still years away from competing in the commercial commodity meat space, but that hasn't stopped many from wondering just what fate Post's cultured revolution may bring for the sector. Animal agriculture isn't just big business; it's a global economic powerhouse, and it will likely undergo a massive overhaul if people like Post and Verstrate are successful. In addition to the sprawling empires of the companies producing, raising, and slaughtering animals, a large portion of crop agriculture exists to feed those billions of animals, too. If clean meat starts taking over a portion of the meat market, huge tracts of land in the American Midwest will look a lot different as feedlots empty and slaughter plants shut their doors for good.

Simply put, meat production is labor-intensive. According to the US Department of Agriculture, a third of US food manufacturing jobs are in meat and poultry slaughter and processing plants, and another nine percent are in dairy manufacturing. Of course, clean meat and dairy products will still need to be processed, but ending the raising and slaughtering of whole animals will likely include efficiency gains in how much manual labor is needed.

There's a lot of money at stake if Post's work becomes commercialized and does well. In the United States, the agriculture lobby is among the most influential forces in the halls of Congress and state legislatures. According to the Center for Responsive Politics, agribusiness interests pour about $130 million into federal lobbying annually, just about the same amount as defense-related lobbying, and far more than labor and trial lawyers combined. And that ag lobby is hardly eager to usher in the "postanimal bioeconomy" that clean animal product enthusiasts envision.

We won't know how the introduction of clean meat will affect the food industry until it's actually brought to market, but it's safe to say that its efficiency compared to today's animal-ag system could leave a huge hole in the ag sector. We'll need much less corn and soybean production, fewer factory farms and transport trailers, vastly fewer pharmaceuticals than are currently fed to farm animals, and of course fewer slaughter plants. For animal welfare, environmental, and public health advocates, these are all huge reasons to be enthusiastic about clean meat. For those who currently make their living by turning crops into farm animals and farm animals into meat, it means huge changes.

A 2013 report commissioned by the United Soybean Board found that animal ag has an enormous impact on the American economy. If you're wondering why the soybean producers are concerned about the fate of animal agriculture, consider that America's biggest buyer of soybeans isn't the tofu industry; it's animal agriculture for feed. Ironically, the last thing soy producers want is for Americans to shift from meat to soy products like tofu and edamame, since the latter require so much less soy. As the Soybean Board report notes, "actions to maintain and expand animal agriculture in the United States—by supporting its long-term competitiveness—are of critical importance to the soybean sector."

When all is said and done, the board concluded that animal agriculture in the United States provides 1.8 million jobs, adds $346 billion to national economic output, and results in $15 billion in income taxes and another $6 billion in property taxes each year. In other words, it's not a trivial matter to replace all those hatcheries, farms, transport trucks, and slaughterers with lab-coated microbiologists and meat breweries.

In a very real way, the success of the clean-meat sector could upend our food industry more dramatically than perhaps any other innovation ever has.

So far, the animal-ag industry hasn't felt especially threatened by Post or his colleagues in the nascent cultured-animal-product world. As a National Cattlemen's Beef Association spokesperson told CNN after the London burger tasting, "We feel confident that consumers will continue to trust and prefer traditionally raised (not lab-engineered) beef. No laboratory product will ever be able to take the place of cattlemen and women or the dedication they have to the customer, the consumer, or to rural America."

The Animal Agriculture Alliance feels the same way. After condemning Post's burger for its enormous price tag, the lobby group's spokesperson went on to parody its desirability. "I think it's safe to say that the cultured meat scientists have some hurdles to overcome, least of all taste. While none of the official testers spit the meat out, no one raved about the 'furger's' (fake burger, get it?) meaty taste and texture. In fact, one tester called it 'surprisingly crunchy.' That's never good." The alliance spokesperson went on to deem Post's burgers "Frankenburgers" and concluded, "Why settle for second best when you can have the real thing?"

And the American Meat Institute joined its barnyard comrades, with its spokesperson declaring that consumers are more interested in meat from their local farm than from their local lab. "A laboratory grown meat product derived from stem cells is unlikely to satisfy the trends currently at play," she assured her industry.

The animal ag reactions to Post's burger tasting bring to mind the adage that "first they ignore you, then they laugh at you, then they fight you, and then you win." At the point of tasting, it seems, the groups representing conventional animal ag moved from the ignoring phase to the ridicule phase. So far, however, there's been essentially no fight from the meat industry at all. If anything, some players in the meat market seem more eager to join the cultured world than to fight it. But many in the conventional meat industry still don't seem to have changed much at all and instead are simply plodding along, seemingly unaware of the storm people like Post have brewing for their field.

In some ways, the conventional meat industry seems at least somewhat analogous to the natural ice industry of old. During the first half of the nineteenth century in America, in-home ice had grown from a rare commodity to a substantial industry. Naturally occurring ice was big business: large cuts of it were harvested from northern lakes and shipped to icehouses where consumers could buy small blocks to store in their home iceboxes, primarily to keep meat and produce fresh for longer.

Enter the invention of industrial refrigeration, and all of a sudden there was a much cheaper way to stock those icehouses with blocks for consumers. Rather than harvesting naturally occurring ice, industrialists could simply cool water down to make their own and then sell to the public. By World War I, the natural ice industry was essentially over but not without a fight. Largely starting in the 1880s, as historian Jonathan Rees explains in his book *Refrigeration Nation*, the natural ice producers defended their enterprise by railing against "artificial ice," warning consumers that the ammonia used in refrigeration could leak into water, contaminating their ice. You don't want that ice touching your food, and certainly not in your drink, they alerted consumers. The irony at the time was that natural ice was actually *less* safe, due to water pollution from local industry and manure from the horses used to drag the ice out of the lakes. So-called artificial ice came from water that was boiled or otherwise filtered for consumer protection. And a century later, virtually all ice we use at home now comes from water we've cooled in our own personal artificial ice-makers that we call freezers, with essentially none of us thinking that there's anything unnatural about it at all. In fact, we wouldn't even consider living in a home without one.

Whether clean meat follows a similar path as "artificial ice," the farm lobby may not be concerned while the industry is still theoretical. After all, even though advances have been made since Post and Verstrate debuted their burger in 2013, the industry still has some work to do to perfect the technology and, critically, bring the costs of production down to the point at which consumers can actually afford their products. But if the clean-meat industry can find a way to bring its meat, leather, milk, and eggs to market, the old guards will have to start paying more attention.

For their part, many of those working in cellular agriculture would welcome support from major food companies. "I want to make sure our new cultured companies don't become the electric car," Isha Datar of New Harvest notes, recalling the role the auto and oil industries played in destroying the embryonic electric car industry in the mid-1990s. "Instead, I want Tyson Foods and other meat companies to be a part of this movement." The Good Food Institute's executive director, Bruce Friedrich, agrees, asking, "Who better to create the best chicken nuggets without chickens than Perdue? Who better to create real Spam without pigs than Hormel?"

So far, Hormel hasn't rushed to embrace clean-meat technologies, but at the end of 2016, signs began pointing in the right direction. Tyson Foods, the largest meat producer on the planet, announced the creation of a new venture capital fund, backed by $150 million of the behemoth's money, to invest in, among other things, alternative proteins. According to a story about the fund in the *Wall Street Jour-*

nal, Tyson intends to invest not just in plant-based protein but also in "meat grown from self-reproducing animal cells [and] 3-D-printed meat." And at a Future Food Tech panel in New York City in 2017, the head of Tyson's fund, Mary Kay James, confirmed the company's active interest in clean meat.

Later in 2017, Israel's meat industry began taking an active interest as well. Eli Soglowek, chairman and CEO of Israel's largest processed meat producer, Soglowek Group, emailed me just before he spoke at Israel's first-ever cultured-meat conference. Organized by the Modern Agriculture Foundation, essentially an Israeli counterpart to the Good Food Institute, the conference attracted investors, entrepreneurs, and scientists from around the globe. Having a major meat producer speaking at the conference was a big deal. And what Soglowek casually wrote me may be an even bigger deal: "[We] are always trying to be in the forefront of the meat industry. We believe that cultured meat would be used in our manufacturing in the next ten years as it would become commercially feasible and affordable."

Just a couple months later, what industry detractors of clean meat imagined just a few years earlier to be unthinkable became all too real. In a historic announcement, agribusiness giant Cargill became the first conventional meat company to announce an investment in a clean meat start-up. Touting its support of Memphis Meats, Sonya Roberts, president of growth ventures at Cargill Protein, remarked that the company believes "consumers will continue to crave meat, and we aim to bring it to the table, as sustainably and cost-effectively as we can. Cultured meats and conventionally produced meats will both play a role in meeting that demand."

Forward-thinking companies like Cargill, Soglowek, and Tyson might be taking a page from filmmaker Canon's playbook. As a 2006 *USA Today* story pointed out, "In the pre-digital age of photography, no brand was more synonymous with imaging than Eastman Kodak. . . . Now, it's in a heated battle with Japanese-owned Canon

for market dominance—and by many measures, Canon is ahead."
As Canon and Kodak were battling each other for supremacy in
the camera industry, the emerging digital era threatened to change
everything, including shuttering high school dark rooms, closing film
developing stations at local pharmacies, and putting the makers of
gelatin film out of business altogether. Rather than keep pace with
the new technology, Kodak lagged, even as Canon dove headfirst into
the digital camera sector.

The result couldn't have been more predictable, and if you're paying attention, you already know: Canon is the leading digital camera
brand, while Kodak filed for bankruptcy in 2012.

This story provides stark contrasts with regard to how major
companies can respond to innovative technologies, like Mark Post's,
that could be transformative to their industry. "Disruption" may be
the buzzword in Silicon Valley, but some established food conglom-
erates seem willing to do quite a lot to maintain the status quo. Assuredly some other big meat companies will join Cargill in expanding
their portfolios early in the game to include clean animal products,
while others will follow in Kodak's footsteps.

Already some major investors are sounding alarms that indicate
an increasing realization in the financial community about what kind
of impact clean meat may entail. In late 2017, for example, Paul Cua-
trecasas, CEO of investment banking firm Aquaa Partners, published
a dire warning in a feedstock trade publication. "Feed companies
should hedge their risk by investing in lab meat," the executive counseled. "If even half as many people move to lab meat as have moved to
synthetic wool, the meat industry would likely face significant bank-
ruptcies."

Tyson hasn't made any actual investment in a clean-meat company yet (though it has made investments in the plant-based meat
space), but in late 2016, it made its first-ever public commentary
on the topic. Sitting on a panel about cultured-animal products at a

biotech conference in North Carolina, the company's lead R-and-D scientist, Dr. Hultz Smith, said, while he doesn't anticipate that traditional animal production will go away anytime soon, he does believe that cell culture can offer consumers another choice for how they get their protein.

For all the talk among those in the cellular-agriculture industry about how their technology is going to render the industrial animal-ag complex obsolete, Smith's prediction will likely end up being more accurate—at least in the near term. Even after clean meat hits the market, there'd surely be people who want to continue eating meat produced the "old-fashioned way" (whether today's factory farming system can really be considered "traditional" seems suspect). But as billions more people populate the planet in the coming decades, it seems pretty safe to say that a good portion of them are likely going to be quite happy to eat meat that was produced so much more efficiently and humanely, especially if it's cost-competitive with conventional meat.

If these technologies actually start displacing (if not replacing) animal ag, there's no doubt that there'll be a major shift in the farm economy. And while the trend for decades has already been a reduction in the number of farmers in America, many farmers and others involved with animal ag will find themselves in need of new jobs. Jason Matheny's prediction that "future farmers of America are microbiologists rather than cattle ranchers" just may turn out to be true.

Paul Mozdziak, a North Carolina State University poultry scientist culturing avian cells with New Harvest funding, isn't overly concerned on this point. He was on the same panel as Tyson Foods' Smith and agrees. Mozdziak predicts that chicken farmers won't need to worry about his cell-cultured poultry meat competing with their birds anytime soon. "But even if that happens," he notes, "farmers are only one percent of Americans, and only a fraction of them are involved in animal ag. The same thing happened to the tobacco farmers all around me. As people stopped smoking, they switched to other crops. From what I can tell, a lot of them are growing sweet potatoes now."

It turns out that a lot of them are also growing chickpeas. A 2013 *Wall Street Journal* story reported how "hummus is conquering America," and the result is that there's a huge new demand for chickpeas that didn't exist before. As the need for tobacco growers declines along with the demand for their product, Big Hummus—represented largely by Sabra, now partially owned by PepsiCo—has been pressing those tobacco growers to start planting chickpeas instead. And many of them are. Farmers adapt to new market conditions, just as people from all professions have to in a free economy.

Tastes change, and so must those providing the foods to satisfy those tastes. And it's of course not only in the food industry that consumers have welcomed dramatic change. How many of us regret that travel agents were displaced by Expedia? Are many tears shed while Netflix-bingeing for the lack of Blockbuster video stores in our world anymore? These are all jobs that we at one time needed, which were lost when superior alternative industries arose to take their place.

Mozdziak points out that farmers have always had to change with the demand, and that the work of researchers like Mark Post and himself aren't much different from other agricultural efficiency innovations. "The land's not going away. If they're not producing chickens or corn, they'll produce something else. And keep in mind, these cultured meat plants are going to have to employ people, too, creating a whole new sector of jobs. That's just the way our economy works."

Yet to be fair, there's a key difference between such shifts. Tobacco growers can learn to start growing chickpeas, but will farmers become microbiologists or tissue engineers? Yes, these products will require farmed nutrients to feed their growing cells, but one of the reasons cellular ag requires so much less land than conventional ag is that you just don't need nearly as much farming to bring these products into existence. What the cellular-ag proponents tout as their great efficiency would almost certainly further reduce the percentage of the population that's engaged in farming—a trend that's been going on for the past century to the point where today 99 percent of Americans aren't involved in commercial farming at all.

But at least for now, most animal ag-vocates seemingly aren't worrying. A few see an opportunity for the industry to get involved in cultured-animal products, but most I've spoken with simply aren't that excited about the issue, and rightly or wrongly, they're not shaking in their cowboy boots.

Dan Murphy, executive director of the US Meat Industry Hall of Fame, doesn't mince his words about people like Mark Post and Peter Verstrate. "Their over-the-top statements about how animal-free shamburgers will shortly revolutionize the world's eating habits are akin to the pronouncements of scientists at the beginning of the Atomic Age about how nuclear power would produce electricity so cheaply that in the future, we wouldn't even need to meter it."

———————

Despite all the attention on Post and Verstrate's cultured burger, several things need to happen before its commercialization becomes a real possibility. First, there's cost. In 2015, two years after the burger tasting, Post reflected on his efforts to bring his costs down. "It was $330,000 when we first publicized the patty. At this point we've already managed to cut the cost by almost eighty percent. I don't think it will be long before we hit our goal of sixty-five to seventy dollars per kilo." In other words, about $11 per burger by 2020, by his estimation, and he's certain they can eventually make it more affordable than conventional beef.

Moving clean meat from the university and to a commercial reality is extremely important to Post, so in mid-2016, he and Verstrate started a business to accompany their academic efforts. Mosa

Meat—named after the Latin word for Maastricht's picturesque Maas River—exists not so much to sell clean meat to the public as it does to serve as an intellectual property–licensing company that will sell its technological processes to companies that want to produce such meat themselves. Verstrate serves as the company's CEO and is already attracting venture capital, including from an undisclosed "meat industry investor." He believes Mosa Meat will sell a small amount of meat to local Dutch consumers, but the company's primary revenue will come from licensing. "Being a large meat producer is not our ambition, and—more important than that—spreading the technology will go much faster if you sell licenses."

(Post has also since started a separate company on his own, Qorium, focused on producing lab-grown leather.)

Verstrate predicts that by 2020, they'll know how to produce the meat at a cost low enough that they can start investing in the actual equipment needed to help their technology into the market. Predicting it may take one or two years after that to actually create a product, he expects, the meat could be available for sale by perhaps 2021.

The fact that Post and Verstrate are ready to convert from academia to entrepreneurship raises questions about the argument that we're too early in scientific discovery for the commercialization of clean meat: Is clean meat science advanced enough now to warrant actual companies, or are dollars better off spent on open-source academic research? As you'll see in this book, several entrepreneurs believe that the science will only advance quickly if the research moves into the private sector.

The Good Food Institute's Bruce Friedrich agrees with the entrepreneurs, arguing that "cellular agriculture has been exclusively the province of the academy for more than a decade, and it seems to me that leaving it there will guarantee that we don't have a product on the market for at least another decade."

To Friedrich's mind, there's ample reason to suspect the sector needs private companies like Mosa Meat and its even newer competitors, and it needs them now. For starters, there are resources that will flow to a private venture that won't flow to a public one. If there were no private companies, the movement would be leaving millions of dollars on the table; millions of dollars that could otherwise be developing this technology. Few of the venture capitalists we'll meet later in this book who are funding these start-ups would otherwise have been offering university grants.

Equally important is whether many of the people in the for-profit tissue-engineering world who work for corporate salaries would even be willing to work in an academic institution on a grant. In other words, one would presume the best tissue engineers in the world know that they can command much higher paychecks at companies rather than at schools. Some of the best will be willing to work in academia, but many likely won't, and their absence from the field could slow the commercialization of clean meat considerably.

Even with their research still proceeding, Post admits that they're far away from creating anything other than ground meat like hamburgers, hot dogs, meatballs, and chicken nuggets. Whole meats made of thicker tissue, like a T-bone steak, are still at this time beyond the grasp of the clean-meat movement. The problem is simply that in the absence of blood vessels, which carry necessary nutrients throughout muscle tissue, those unfortunate inner muscle cells in culture are deprived of their sustenance and perish. One solution Post believes may solve this problem involves 3-D printing tubes in between the tissues to transport nutrients. For now, though, given the large quantity of meats that are sold ground, Post is content to work on this lower-hanging fruit, or lower-hanging meat, if you will.

Even getting the right type of bioreactor to grow these cells at scale will be a challenge. Right now such equipment is only used for medical purposes, where cost is less of an issue and the size is much smaller. Entirely new generations of bioreactors are needed to be invented to bring clean meat to the market. Normally a bioreactor is used for cells that can swim in liquid suspension. There just aren't yet any large bioreactors used to engineer solid tissue. Post theorized that perhaps microcarrier beads in the bioreactor could allow muscle cells to grow on them, leading him to actively start experimenting with different kinds of such beads.

"The purpose of the unveiling in 2013 was to tell everyone that this is possible," Post notes. "That goal has been reached. The goal of the next unveiling will be to have a commercial product that's already been tasted and approved by the food and drug administrations of several countries."

The big question for Post isn't whether they'll commercialize it—of that he's certain. The question is whether meat-eaters will eat it.

"I always ask audiences, would you eat a hot dog? And nearly all of them laugh and say yes. Then I ask: Do you know what it's in it? And they all say no." Post thinks that if consumers honestly evaluate the merits of clean meat compared to conventional, they'd switch over in droves. "People eat things all the time they don't know anything about, and in this case, the more they know, the better they'll feel about it. Right now they're eating meat from animals who were doped up with all types of drugs and lived in terrible conditions. Why wouldn't they want to switch to something so much cleaner?"

Driven by the same desire that fuels Jason Matheny, Post gets serious about the "coming meat crisis" he views as inevitable with rising global population and income. The only question for him is whether humanity will rise to the occasion and invent a solution—that they will be willing to, as Sergey Brin hopes, try something new—before it's too late.

"My goal is to replace the entirety of meat production with cultured meat," Post boldly proclaims. "In fact, I think that once we have suitable alternatives to it, we'll come to think of slaughtering animals for food as indefensible and it will be banned. Simply put," he predicts, "the meat industry as we know it has no future."

Verstrate sees things slightly differently, agreeing with his co-founder that factory farming of animals will cease to exist. But just as there's still demand for horse-drawn carriages, whether as a tourist attraction or by religious or other groups that don't adopt technological advances, such as Amish communities, Verstrate predicts there'll still be some specialty meat production coming from the slaughter of animals. "But we won't have anything like we have today."

In 2016, three years after their public testing, and from his perch where that first burger was produced, Post forecasts the future. Standing before me in a white lab coat in front of a wall of incubators and underneath a map of the Netherlands—the only nonmedical sign in the cramped space—he reflects on what his work may unleash.

"Twenty years from now," Post envisages, "if you enter the supermarket, you will have the choice between two products that are identical. One is made in an animal. It now has this label on it that animals have suffered or have been killed for this product. It has an eco tax because it's bad for the environment. And it's exactly the same as an alternative product that's been made in a lab. It tastes the same. It has the same quality. It has the same price or is even cheaper. So what are you going to choose?"

4

LEADING WITH LEATHER

During World War I, Germany had a real problem. Its zeppelins—rigid, cigar-shaped domes held afloat by gas-filled bags—were sowing terror over Britain, but they required vast resources to produce. The scarcest of those resources was the very bags that kept the weaponized vessels in the air.

Germany's storied sausage industry had always used cows' intestinal linings as a by-product, necessary to hold together ground meat. But cow intestines were found to be particularly useful for creating "goldbeater's skin": an ultrathin, lightweight material used to contain the hydrogen or helium that lifted zeppelins toward the heavens. As useful as the cow intestines were to the military, their relatively small size, compared to the magnitude needed, meant they weren't especially efficient. In fact, just *one* zeppelin required the intestinal lining of more than a *quarter million* calves.

With the military need for zeppelins increasing, the demand for bovine intestines went sky-high. Pretty soon, Germany and its allies halted all sausage making in order to ensure the military received as many cow guts as possible. But even with the sausage bans, there just weren't enough cattle to keep the zeppelin fleet afloat.

After Germany's defeat in 1918, production of zeppelins plummeted, and sausage makers could go back to their craft. But even so, the American tire and rubber company Goodyear set to work seeking a better way to fly the aircraft that were still being utilized for non-military purposes. Eventually, they developed a gelatinized rubber that was much cheaper to produce and could prevent future intestinal shortages. In fact, by the 1930s, no German zeppelins contained goldbeater's skin, having all switched to the rubber instead.

This story, set in a far different field, illustrates what Modern Meadow is trying to accomplish. Using a quarter million cows to float one aircraft was simply unsustainable, and Andras Forgacs, whose steak chip I ate in 2014, argues the same is true today for our fashion industry. Relying on animal exploitation to clothe ourselves simply uses too many resources and creates too many environmental problems.

But how do we start replacing animal products in our economy with the equivalent of Goodyear's zeppelin rubber? Mark Post proved to the world that you can grow parts of an animal outside that animal. As revolutionary as that is, and as much promise as it holds to upend the conventional meat industry, the question of just how best to introduce the public to this concept remains an open one, especially given the unanswered questions about how the market will respond to such an innovation.

After all, while the number of animal products that could be replaced through culturing technologies is nearly limitless, one product or industry must be chosen for the initial disruption. Not everyone agrees it ought to be meat. In fact, Forgacs ultimately concluded that, just like one animal-based fabric—goldbeater's skin—was replaced by a superior animal-free version, the cultured sector needs to lead not with meat but by competing with another animal-based fabric: leather.

In addition to the potential marketing problems clean meat could face when it's ready for commercialization, growing leather is technically simpler than growing meat. Unlike meat, which is three-dimensional, skin sits largely in the 2-D world. There'll also be few or no regulatory hurdles to introducing biofabricated leather into the market, which may not be the case for clean meat.

Companies like Mosa Meat are hoping there'll be an appetite for their food when it hits the market several years into the future. But will their meat be more palatable to the public if years earlier we've already begun wearing cultured leather? Getting people comfortable with the concept of cellular ag is a major focus of Modern Meadow, and it intends to do it by focusing its efforts not on food but on fashion. To Forgacs, biofabrication of animal fabrics is the key to ushering in the postanimal bioeconomy.

"Leather is a gateway product for our nascent cultured industry," Forgacs asserts. "I've got all the respect in the world for Mark Post, but the first breakthroughs for cultured-animal products likely won't be in meat. They'll be in leather. Wearing it before eating it will ease consumers into the concept of using animal products produced without the animal."

———

As best we can tell, we've been wearing the outer portion of other animals for at least tens of thousands of years, and probably longer. Ever since *Homo sapiens* started spreading out of the warm climate of Northern Africa some one hundred thousand years ago, clothing went from being largely decorative and sparse to downright essential for survival. Even *Homo neanderthalensis* (known to us popularly as Neanderthals, the now-extinct human species living in Europe for hundreds of thousands of years before we *Homo sapiens* showed up on the continent) were covering their bodies with animal furs to keep warm in the northern climates. One 2012 study found that our Neanderthal cousins would've needed to cover up to 80 percent of their bodies with fur to survive the European winters of their time. In other words, human brains didn't wait until our bodies evolved to endure colder climates; they simply figured out how to procure external sources of warmth from the animals who had already evolved to survive there.

Today, most of us don't live in the equatorial climate from which we originated, and our bodies alone aren't sufficient to keep us thermoregulated at latitudes in which we didn't evolve. Even those of us who do live in tropical regions today still (obviously) prefer wearing clothing, and often that clothing comes from animals. Yes, we've found all types of plant-based and, more recently, synthetic ways to cover and warm ourselves, but one animal-based fabric still dominates a lot of our clothing choices, especially when it comes to items like shoes, bags, and more: the skins of cattle.

Fur-based clothing is truly ancient, far predating civilization by many tens of millennia. But once we were settled into communities, learning to tan animal hides soon followed. Perhaps the earliest evidence of humanity's use of leather dates back to the Neolithic period, some five thousand years ago, when people in what's now Armenia had apparently figured out how to tan animal hides to turn them into leather shoes. The Egyptians three thousand years ago had less use for leather as clothing, though they appear to have used it more regularly for furniture and bags, and even dog collars. But there may not have ever been a time when leather was as important to our species as it is today, making it riper than ever for a major disruption.

The American leather industry is a global behemoth. In exports alone, the US cattle sector sells $3 billion in hides each year from the thirty-five million cattle who end up in American slaughter plants.

About half of American leather use goes to shoes, a third to furniture and car seats, and the rest to accessories. Globally, the market for leather is valued at more than $100 billion. (Many other animal skins are still quite popular, too. For perspective, the global animal fur industry is valued at around a whopping $40 billion, and python skins alone—in other words, not including other reptile skins like those from alligators and crocodiles—account for about a billion dollars in global sales.) But in addition to the problems associated with the animal-agriculture industry that produces all these cattle, there are other problems with how their hides get turned into leather.

If you've ever wondered why your leather watchband doesn't biodegrade despite just being dead skin, it's because of tanning—a type of mummification that ends up preventing the skin from rotting right off your wrist. The process is multistage and requires many different chemicals, which may be one reason it took humans so long to go from fur coats to leather shoes. Tanning essentially binds up the collagen in the cow's skin by stabilizing it, permanently altering the protein structure of the hide and ensuring a far longer life span than it would otherwise have.

But tanning can be damaging to the environment, workers, and communities surrounding the tanneries. As *Quartz* observed in a 2017 profile on Modern Meadow, "Traditional leather production leaves behind a vast carbon footprint, a destructive trail of environmental pollution, brutal animal suffering, and, often, disturbing human-rights violations."

First, the skin needs to be chemically stripped of hair, fat, and other undesirable content left over from the butchering process using a harsh lime. This waste—chemicals and all—can go straight into the trash. The next step in the modern tanning process typically involves soaking the skins in a vat of chromium, a caustic substance that then can get dumped into local waterways, especially in nations with lax environmental laws, such as the big leather tanning countries of India and Bangladesh. In the latter, officials essentially admit they don't enforce environmental rules in the major tanning center of Hazaribagh. As a result, toxic tanning chemicals like chromium sulfate, sulfuric acid, and others routinely get dumped—untreated—into local waterways. Workers there—including children—are subjected to these dangerous chemicals with often no or little protection. As one of them explained to international charity Human Rights Watch, "When I'm hungry, acid doesn't matter—I have to eat."

In India's leather capital of Kanpur, tannery pollution of the Ganges River became so terrible that the government was forced to shut down in excess of one hundred of the worst offenders in 2009—more than a quarter of all tanneries in the region. The polluted water is associated with higher rates for the local population of skin problems, respiratory illness, renal failure, and even blue baby syndrome.

Wildlife, particularly aquatic animals, can be especially hard-hit by the process. The waterways near the tanneries often become dead zones where, as the name implies, no animals can survive. And the toll on the workers themselves—most of whom are among the poorest in the world—can also be dramatic. Chromium exposure is linked to a host of ailments prevalent among tannery workers. From asthma and bronchitis to lung cancer, the list of increased workplace hazards associated with tanneries reads like the rapid-fire disclaimer language at the end of a prescription-drug commercial. A particularly painful problem for tannery workers is the prevalence of "chrome holes," what used to be referred to in medical journals as "tanner's ulcers," that can develop on workers' hands and even their nasal passages. The circular wounds appear like canker sores, but on the parts of the bodies most often exposed to the chromium, either directly from touch or by inhalation.

Knowing this, it's easy to see why producing leather in a lab, rather than removing it from an animal, is so appealing. When Modern Meadow makes its leather, it has no need to produce hair, flesh, or fat, so the tanning process is shortened just to the secondary stages. "The whole process is a lot shorter," Forgacs notes, "and there are fewer steps with much less effluent." In short, Modern Meadow is only practicing the final stages of tanning, essentially preserving the skin and then treating it to achieve the required thickness.

And since he doesn't need to worry about government regulations for disposing of chemicals conventionally used in the first stages of tanning—since he's not using them—Forgacs plans on tanning his hides in the United States, possibly even in New York City, where the company is based. Keeping production close to home will also minimize transportation costs usually associated with the sprawling global leather market, including that from where most of our imported leather imports originate: Asia.

The benefits of reforming the tanning sector seem obvious. But will wearing lab-grown leather really make us more likely to eat similarly produced meat? Consider the following thought experiment: Imagine how you'd feel about eating a burger that was grown outside a cow. Even if you're on board with the concept of clean meat, like I was at first, you might be a little hesitant about taking your first bite. But what if you were presented the opportunity to wear a pair of shoes produced with real leather that was produced in a lab? If you're like every person I've queried on the topic, you have essentially no qualms at all about the latter. Wearing a novel material just doesn't seem odd to many of us, or at least not as odd as eating a novel food. (Though, to be fair, a lot of the food sold in grocery stores today is pretty novel as far as human history is concerned, and most people are quite content to eat it.) You probably already wear synthetic leather running shoes without even thinking about the fact that they're synthetic. You may not even know if your athletic shoes are real leather or not. (Many of them aren't.)

That's what Andras Forgacs is hoping, and what Modern Meadow is now working on. The creator of the steak chip has put his company's early experiments with meat on ice and is focusing exclusively on the leather market in the hopes that he'll be able to hasten consumer acceptance of cellular agriculture by producing a wearable product instead of an edible one.

Forgacs got the idea while living in China in 2011. At the time, he was working with his father to advance a biomedical company they cofounded four years prior. While at Organovo, they developed a 3-D bioprinting process that produced real human tissues in vitro, giving drug companies the opportunity to test new compounds without having to use organs from an actual person. Forgacs fortuitously found himself in conversation with a leather industry executive who helped move the gears in his brain to conceive of a new concept altogether. "Since you can grow human skin," the CEO pressed him, "could you also grow a cow's skin? Think how much money I could save on shoes if you could just grow me the leather," the businessman fantasized. "Why grow the whole cow when all I want is the skin?"

Forgacs had never given much thought to this idea. Yet after conversation with other leather industry leaders, he began to think that maybe lab-grown really was a feasible business idea. "The entrepreneur in me said yes," Forgacs recalls, "but honestly I wasn't really sure."

Still, he wondered about the possibility. If we can make medical-grade human tissues that can be fully functional, shouldn't it be possible to make food or even textile-grade animal tissues? The question also appealed to his desire to apply scientific breakthroughs to solve the serious sustainability problems involved with meat production.

That idea became the basis for a new start-up focused on growing animal cells for both meat and leather. As a result, in 2011, two years prior to Mark Post's burger tasting in London, Modern Meadow, the world's first cultured-animal-products company, came into existence.

Today, huge amounts of leather get wasted because cows stubbornly don't come in shapes resembling wallets, shoes, and wristwatches, but in the lab, you can grow leather into any shape you want. When you divorce the cow from the process, you enable new design and performance possibilities, including the ability to create leather as thick or thin as you want, as light or heavy as you want, and even translucent or other totally new types of leather we've never had. "This isn't about leather imitation," Forgacs informs me. "It's about leather innovation."

Part of Modern Meadow's plan to bring its goods to market is to maximize the advantage its stability can offer clients in the fashion industry. The strategy isn't so much to produce its own shoes and jackets, but rather to supply the raw material to leather goods makers who will then turn it into handbags and other leather products. No longer will these manufacturers be subject to the price fluctuations of the meat market. If a drought hits and the price of feed crops skyrockets, leather goods manufacturers won't have to worry. If there's an outbreak of a disease like mad cow that causes cattle to be condemned en masse, there's no disruption in Modern Meadow's process. In very real ways, removing the leather industry from the cattle industry could be a dream come true for leather manufacturers. And it could also be a dream come true for the planet.

———————

Like Post, Forgacs isn't a vegetarian, and also like Post, his original funding came not from a billionaire investor—though that would come—but rather from his own government. Forgacs applied for and received research grants from the US Department of Agriculture (USDA) along with the National Science Foundation (NSF) to help launch Modern Meadow in 2011. The USDA wanted to fund Forgacs's meat research, while the NSF sponsored his experiments with leather.

As he was bringing in six-figure grants from federal agencies,

Forgacs also sought out private money, leading him to Breakout Labs, the grant-making arm of PayPal founder and billionaire Peter Thiel's venture capital empire. And as he had with the USDA and NSF, Forgacs persuaded Thiel's foundation that his young company was promising enough to yield a $350,000 investment. "Our focus is on funding breakout science, especially ideas and inventions that have the potential to literally change the way human beings live," says Lindy Fishburne, the executive director of Breakout Labs and senior vice president of investments at the Thiel Foundation. "Ask yourself how many major industries would be impacted if Modern Meadow successfully delivers biofabrication of leather."

After Thiel's six-figure influx, Forgacs began attracting venture capitalist money from many of the biggest names in Silicon Valley, including firms that were early investors in some of the leading names in tech, like YouTube, SurveyMonkey, Apple, Oracle, and PayPal. He even pulled in funding from Justin Rockefeller, the great-great-grandson of the oil baron John D. Rockefeller.

At first focused on both the inside and outside of the cow (beef and leather), pretty quickly Forgacs had a change of heart. Steak chips were certainly much easier to produce than steak, and he really did love the idea of consumers picking up a bag of his chips. But there were still a lot of problems. Despite the positive initial reaction to the steak chips from investors, Forgacs contemplated the serious barriers to commercializing his meat. "People have really strong opinions about food, especially when it comes to new technologies," Forgacs notes. "They have less-strong opinions around new materials like GORE-TEX and carbon fiber."

As some in the field have pointed out, the clean-meat industry, in some ways, still lacks the "right story" to sell its novel food. In other words, it's not enough to state the facts about the new technology; you have to introduce people to it in a way that makes them feel comfortable.

By way of example, prior to the 1970s, Americans really just weren't that interested in eating sushi. Between the seaweed and the raw fish, the dish was simply too foreign and exotic for American palates of the time. The story of how it became a favorite of the American diet is perhaps apocryphal but still relevant. An innovative Japanese chef in Los Angeles, the story goes, wanted to entice his clients to start eating more sushi. Knowing that Americans weren't exactly enthusiastic about eating seaweed, he turned the sushi roll inside out, putting the rice on the outside and seaweed on the inside. Next he replaced the fatty raw tuna with an equally fatty plant-based food Southern Californians already loved: avocado. The final step to gain acceptance was to stop using Japanese names for the roll. And thus was born the California roll. Sales took off, and Americans eased their way into eating this culturally unusual food, eventually learning to love all types of sushi.

That may be just what the cell-ag community needs: some type of entry product that eases people into the idea of growing animal products. Once consumers are accustomed to the idea of wearing lab-grown leather, will the idea of eating lab-grown meat be as foreign as it may appear to some today? It certainly seems possible that leather could pave the way for the beef, and ultimately other meats, too. The sushi story may not be a perfect analogy (after all, Americans were already accustomed to eating rice, fish, and vegetables, even if not all together), but it does offer a valuable insight into the problem facing the purveyors of animal-free animal products.

Since people seemingly have no problem wearing synthetic leather all the time now, it's likely that Forgacs is correct when he predicts no one will bat an eye at wearing Modern Meadow's real leather when it hits the market. But whether that makes people more likely to eat clean beef or not is still an open question. There's a big gap between what we're willing to put on our bodies as opposed to in our mouths, and it's not as if synthetic leather's acceptance has changed how we feel about synthetic food (though most of us regularly eat foods with synthetically produced ingredients, such as flavorings or rennet in cheese, even if we don't think about it).

In a way, the trail for biofabricated animal fabrics is already at least somewhat blazed for Modern Meadow. Unlike with clean meat, some people are already beginning to buy lab-grown animal-based garments, many of which utilize comparable technologies to those employed by some of the companies discussed in this book. For example, California-based Bolt Threads is growing in vitro spider silk (what their webs are made of), starting with yeast cells that've been engineered to spit out the proteins naturally found in the extremely durable arachnid product. Unlike the more common silk from worms—who've been domesticated and bred for silk production over the course of many centuries—spider silk is far stronger, some types being even sturdier than Kevlar, all the while being as soft as, well, silk. The problem with trying to produce it commercially is that spiders don't do so well when we try to farm them, typically eating one another in the crowded conditions needed for insect farming to work. Cannibalism just doesn't lend itself to profitability. (A team in Madagascar did succeed in producing a farmed spider silk garment in 2009, but only after four years of farming *a lot* of spiders.)

With $90 million in venture capital raised, in 2017 Bolt Threads announced its first commercial product—a necktie that retails for $314 (as math fans they're big on pi, or 3.14), and were only made available to fifty lucky individuals who won a lottery to buy them. The company also inked a deal with Patagonia for its arachnid-free spider silk garments. A Japanese competitor named Spiber (as in "spider fiber") is doing the same thing and in 2015 partnered with North Face to produce the so-called Moon Parka, a durable winter coat containing their lab-grown silk that is, at the time of this writing, available for sale in Japan and retails for $1,000. And shoemaker Adidas is already starting to use lab-produced spider silk, called Biosteel, manufactured by a German competitor of Spiber named AMSilk. The company boasts that "a spiderweb made of pencil-thick spider silk fibers can catch a fully loaded Jumbo Jet Boeing 747 with a weight of 380 tons."

But it's unclear whether the new lab-grown spider silk clothing will have much influence on how consumers end up feeling about Modern Meadow's lab-grown leather. Aside from the fact that many people wear leather daily (unlike silk, and *especially* spider silk), a sheet of cow's skin is more complex to produce than strings of spiderwebs, and leather is very easily recognized visually by consumers as leather, also unlike silk.

When Forgacs first began culturing leather, he used a process similar to what other clean meat companies are doing: take a biopsy (this time of skin instead of muscle cells), allow them to grow and produce more collagen, spread it out to form sheets, and then layer the sheets on top of one another. As he continued perfecting his biofabrication technique, the process became more and more refined. Recognizing that the most important part of the leather is simply collagen, Forgacs figured, why not ditch the cow cells and just grow collagen on its own, since it can hold everything together? The very word "collagen" comes from *kolla*, Greek for "glue." And with that, the first-ever real cow-free cow leather was being made without even any starter bovine biopsy at all. (We'll discuss this method, known as *acellular* agriculture, in chapter 7.)

When you're building the leather from the collagen molecule up, the possibilities are essentially endless. Forgacs is producing bovine collagen but is quick to point out that with a simple change in the amino acid sequence, he could be producing crocodile or alligator collagen, too. In a way, it almost doesn't even make sense to think about the skin as belonging to any one animal species, since the platform is so malleable one could create all types of "skin," including skins evolution hasn't even produced. Modern Meadow's chief creative officer,

Suzanne Lee, made the point in a late 2016 interview with *Forbes*, "If you're not bound by the animal, then you can construct it in the way you want. Collagen really is the material we're producing—we can form it in new ways and create leathers that couldn't exist in nature."

In addition to the huge efficiency and functionality advantages Forgacs believes his leather will have, he also notes that it's the highest-value density product he could go after. Leather typically makes up about 10 percent of the economic value of a cow entering a slaughter plant, with the animal's insides—especially the muscle—making up the rest. But on a per ounce basis, the leather is more valuable than the muscle, meaning that it will be easier for Modern Meadow to compete on price than it would if it was focused on commercializing meat.

Forgacs believes his advantages in that competition will start right out of the gate. Presently, between 30 to 50 percent of cowhide gets tossed into landfills or is used as extremely low-end filler material. Whether because it's the wrong shape, or because it has imperfections like scars, insect bites, and other problems, there are a lot of ways a hide's value decreases while on the back of a cow. In the lab, though, the skin is pristine and will of course only get produced in the shapes needed.

————————

Back in 2013, before Modern Meadow had moved from its birthplace in Missouri to the Brooklyn Army Terminal to be closer to the fashion industry, Andras and his father, Gabor, had a meeting with Josh Balk, cofounder of the food technology company Hampton Creek and a high-profile staff member at the Humane Society of the United States. Balk was working with more and more start-ups seeking to compete against the factory-farming model of food production and had been introduced to the Forgacs family by Jason Matheny, who knew he'd be impressed by Modern Meadow.

In 2014, the year after his initial meeting with the Forgacs duo, Balk traveled to Hong Kong to meet with Horizons Ventures, a seven-figure investor in Hampton Creek. Over a tea-and-toast breakfast inside one of the city's fanciest skyscrapers, Balk and Horizons cofounder Solina Chau were discussing what other opportunities were arising in the food-tech space when Balk encouraged Chau to support Modern Meadow.

Intrigued by Balk's pitch, Chau brought the idea of investing in the company to Li Ka-shing, the man whose capital fuels Horizons' entire fund. One of Asia's wealthiest men, Mr. Li reportedly enjoys a net worth of $34 billion. He also happens to be a devout Buddhist and near vegan. Named by *Asiaweek* as Asia's most powerful man, Li owns a vast array of businesses, and through Horizons Ventures the octogenarian tries to expand his wealth and influence even further. No wonder he's affectionately referred to in China as "Superman."

In 2014, Superman became the first major investor in a cellular-agriculture company and announced a $10 million pledge to Modern Meadow to support its venture to commercialize lab-grown leather. Such an infusion of cash was easily the single biggest investment in cultured-animal-product research in history. In fact, it was more money than had ever been spent on such efforts than all previous investments from governments and private investors *combined*, bringing about a new dawn in biofabrication. Within months Forgacs would more than double his staff, eventually bringing on David Williamson, one of DuPont's top technology leaders, as Modern Meadow's new chief technology officer.

"We all know that it's going to be several years before cultured meat is on the market," Forgacs observes. "The same with commodity biofabricated leather. But with at least luxury leather products, it's possible we could be there in the near term." In 2016 *Crain's* business journal reported that Forgacs expects his leather to hit stores sometime in 2018, though he hedges a bit when asked about it, saying that he expects to have a demo plant by then, but he feels certain his leather will be on the market "quite soon." Already, Modern Meadow is producing large sheets of leather and is able to rapidly make new iterations, a key marker of their march toward the market.

With the continued technological progress of Modern Meadow, Horizons took an even greater interest in its holding as it saw the real potential for commercialization, and in mid-2016, Horizons led a new round of funding that totaled an additional *$40 million* for Modern Meadow.

Just a month after the unprecedentedly massive funds landed in Modern Meadow's bank account, I stood in Forgacs's Brooklyn conference room with both him and Williamson, where the chief executive proudly handed me a sample of black leather. It was two years later, nearly to the day, that I'd stood in the same office and eaten one of the world's first steak chips. To me, the leather was indistinguishable from an actual cow's leather. Moreover, it only took mere weeks to grow, as opposed to leather from a cow, which takes years.

Flush with cash and new headlines heralding the fund-raising, Forgacs was ready to start the path to commercialization. He's already working with partners in the luxury leather industry to try their hand at making the valuable items with his branded leather. He predicts that he'll primarily be selling first high-end leathers on the upper end of the cost spectrum. "We'll be competing at first on value, not price," he notes, meaning that his first products won't be affordable for the masses and instead will be aimed at high-end fashion consumers. But the plan is to have Modern Meadow's first full-scale facility start generating the most affordable leather products, too, in the immediate years to come.

In a way, what Modern Meadow is doing is somewhat similar to the lab-grown diamond industry. Like cattle production, diamond mining is fraught with ethical and environmental problems, and scientists have now figured out how to produce actual diamonds—

essentially identical to those that are mined—in the lab for 20 to 40 percent cheaper than diamonds from the ground. In some cases, the human-made diamonds are even brighter and more "perfect" (from a jeweler's perspective) than those from mines. It'd be easy to see how those diamonds could be marketed as just conventional diamonds that are simply more affordable. But a quick search for "lab-grown diamonds" shows their marketers are taking full advantage of the benefits their products bring. Terms like "eco-friendly," "conflict-free," and "pure-grown" abound on the lab-diamond websites. And at least one lab diamond-maker has taken to referring to the mined variety as "dirt diamonds." (One wonders how long it'll be before human-made diamond companies start marketing their stones as "clean diamonds.")

The Federal Trade Commission has since stated that terms like "laboratory-grown" and "laboratory-created"—sometimes also referred to as cultured or cultivated diamonds—would "more clearly communicate the nature of the stone." Not exactly the most romantic name for a diamond ring, nor does a diamond grown over the course of a few weeks in a lab seem as precious as one mined from the earth after millions of years of natural formation.

That may be why lab-grown diamonds account for only a tiny sliver of the diamond jewelry industry, even though De Beers, the jewelry giant, is now producing its own cultured diamonds, though for industrial use only, not for fashion. (De Beers is now also selling to jewelers special machines it invented to help them tell the difference between natural and lab-grown diamonds, as even a microscope alone can't detect it.) Since diamonds are important for many non-jewelry purposes, for example in the electronics market, where they're used as heat sinks, and in industrial cutting, which uses them for their strength and precision, there's still ample demand for cultured diamonds. But human rights advocates who decry the conflicts diamond mining is exacerbating in Africa are hopeful that jewelry seekers will become comfortable with, and perhaps even preferential to, lab-produced gems. Already the chance to get functionally equivalent diamonds for so much cheaper than those that are mined is fueling the growth of the human-made diamond market.

The prognosis for lab-grown leather could end up being better than that of diamonds, however. Part of the allure of diamonds is that they're "rare," even if that's more myth than reality. Because we value materials that are perceived to be hard to come by, a lab-grown diamond just doesn't seem that special. Of course, no one thinks of leather as rare, even if some brands of leather products (think Hermès purses) may be highly valuable. Unlike diamonds, few people think they're getting anything that special when they buy a commodity leather belt or wallet. But even if they do only as well as lab-grown diamonds are right now, Forgacs's investors would likely be quite pleased.

While those investors have had the biofabricated leather space to themselves for most of the company's existence, a company started in 2016 in cooperation with the University of California, San Francisco, calling itself VitroLabs, is now attempting to compete with a product it calls Kind Leather. Starting with cells from cattle, ostriches, and Nile crocodiles, VitroLabs asserts it's developed a better method of growing skin than Modern Meadow. Instead of starting with collagen, VitroLabs is starting with what are called induced pluripotent stem cells, essentially stem cells from adult animals (as opposed to embryonic stem cells) that can be fashioned into any type of cell you'd like.

In a letter it circulated to potential investors at the end of 2016, VitroLabs expressed confidence in why its method is superior to Modern Meadow's. The letter mainly focused on the company's assertion—disputed by Forgacs—that VitroLabs' stem cells will help provide a cellular structure absent in Modern Meadow's method that includes the epidermal layer that gives leather its natural texture. VitroLabs' aim is to have preliminary pilot production by 2018 and claims to have produced and tanned two rounds of leather samples as of this writing.

Forgacs welcomes competitors to the field but sees plant-based leather manufacturers, like MycoWorks—a Bay Area start-up making an alternative leather from mushroom spores—as more competitive players in the space. Through processing mycelium—the rootlike fibers of mushrooms—along with plant-based agricultural by-products, MycoWorks is producing materials that look and feel like leather. Of course, unlike Modern Meadow, MycoWorks is still producing an alternative to leather, not actual leather.

But Forgacs' main competition isn't from lab-grown leather or mushrooms. It's the conventional leather production industry, demand for which has eroded to some extent due to cheaper synthetics for decades. Yet just like mined diamonds, until recently the leather producers have never had to compete with an indistinguishable or even functionally superior product. That's all about to change, if Forgacs gets his way.

As Modern Meadow closes in on the holy grail of commercialization, decisions abound. The question, for example, of whether to brand itself with a name that makes it clear to consumers that they're buying something different is a crucial one. It's easy to see the benefit of simply selling biofabricated leather to goods manufacturers and letting them use the product without ever denoting the advantages and in vitro nature of their new line.

After all, how many consumers even think about what material their shoes are made of? That said, given the enormous functional advantages that clean leather can offer, being up front about their techniques might prove to be a marketing advantage, with some consumers specifically seeking out the first products made with this new leather.

Forgacs is leaning toward going all in on the branding. At a New Harvest symposium in 2016 he asked the audience to raise their hands if they could name a brand of leather. No hands went up. Of course, there are popular brands of shoes, or handbags, or wristwatches that are made with leather, but just not that many actual name brands of the leather itself. "It's a hundred-billion-dollar raw-materials market, and in terms of the actual leather, no one is branding it. I want to create a leather brand that's desirable, has sustainable production, and tunable properties."

As to what to actually call this leather brand, the entrepreneur pondered many options after focus groups and other market research. When initially pressed for his favorite ideas, Forgacs smiles as he contemplates his potential brand name options. "I love that Marvel Comics invented 'adamantium,'" he says, referring to the fictitious and virtually indestructible material that coats the X-Man Wolverine's claws and skeleton. "I'm not looking for a name that connotes indestructibility, though. Before settling on our name, we jokingly used the term 'unobtanium' from the *Avatar* movie as a placeholder around the office," Forgacs notes, referring to the rare and extremely valuable substance coveted by the mining company on the fictional moon Pandora. But in the end, he decided to name his new biofabricated leather "Zoa," the Greek word for *life*. "The essence of what we're doing is bringing life to new materials, and Zoa helps tell that story."

Having now raised more than $50 million, Forgacs walks through his office of several dozen employees laboring away. All of them are working on leather, with the meat side of the business relegated to a potential future opportunity. Steak chips will simply have to wait. But even with investor money in hand, he feels like he's still got so far to go and isn't taking the company's initial success as a sign of inevitability.

"The funding is great, but it doesn't guarantee us success. It's like going to the basecamp of a mountain peak and being glad your supplies are there for you. But you still have to do the work to actually climb the mountain, and a lot can go wrong." In other words, they need to get right both the science and the rollout of one of the most high-profile initial cell-ag products that will hit the market.

Williamson, the firm's CTO, compares the company's efforts to the transformation now being seen in the energy industry. "Major investments led to great innovations to help us start weaning ourselves from coal and oil. What we're trying to do is no different."

If Modern Meadow—or competitors like VitroLabs—have the impact their founders intend and if their branded leather will serve as the gateway to broad consumer acceptance of other clean animal products, the benefits for all the cultured companies will be manifold. The major investment from Li Ka-shing gives Modern Meadow huge advantages in that race and has put the company within striking shot of revolutionizing an industry in desperate need of disruption.

Fur-bearing mammals were the first animal-based source of clothing for humanity and remained our primary means of external warmth for much of human history. For the past several millennia we've been more reliant on our exploitation of cattle than of fur-bearers to clothe ourselves. Putting the ethics of such animal use aside, when the human population was much smaller than today, raising those cattle and tanning their hides presented far less threat to our planet and public health. But today, with nearly eight billion of us needing clothing, producing coverings for our bodies the way our ancestors did poses serious threats, which is why the leather industry is so ripe for reinvention.

"I'm convinced that when we look back in thirty years on today, how we raise and slaughter billions of animals to make our hamburgers and handbags, we'll see this as being wasteful, inhumane, and indeed crazy," Forgacs presses. "We need to move past just killing animals as a resource to something more civilized and evolved. Perhaps we're ready for something literally and figuratively more cultured."

5

CLEAN MEAT COMING TO AMERICA

Andras Forgacs and his colleagues at Modern Meadow view leather as the surest entry point into the market for cultured animal products. But the primary reason cattle are raised is for their meat, not their skin. Removing some of the market for their skin by biofabricating leather would make cattle production less financially lucrative, but as long as we have high demand for beef, cattle are going to be raised en masse, taking a substantial toll on the planet.

That is, unless we can produce that beef without the cattle.

It's probable that lab-grown leather will hit the market before beef, at least in a widespread manner. Forgacs is right: It's much less difficult to produce than meat, faces fewer regulatory hurdles, and is an easier pitch to consumers. But the meat of the problem, so to speak, is meat, and someone's got to be the first one eventually to produce it commercially. Mark Post proved it's possible, and Andras Forgacs gained attention for his novelty steak chips. But knowing that real change won't come until the clean-meat industry develops a true alternative to conventional meat and makes it widely available for mass consumption, Dr. Uma Valeti is committed to making this happen.

Valeti's concerns about how humans consume meat goes back to his childhood. When he was twelve years old, living in the southeastern Indian state of Andhra Pradesh, Valeti attended his neighbor's birthday party. Kids gleefully sprinted through the front yard on the picturesque day, their mothers having given up on trying to calm their screaming children for the afternoon. Sweets flowed like water, and the sugar high helped fuel the birthday bliss. But as the merriment carried on in the front yard, the young boy decided to take a break and venture to the back of the house.

The contrast that awaited Valeti could hardly have been starker.

As children just yards away were having the time of their lives, the animals who would soon be feeding those children were terrified for theirs. Valeti saw a goat tied to a stake in the ground trembling as she watched a second goat be held down for slaughter, violently kicking in vain protest for his life. Chickens in cages sat paralyzed and they, too, one by one, joined that goat on the butchering block.

Over the cries of the doomed beasts, Valeti could hear the cheerful families in the front singing "Happy Birthday."

"That's when it really struck me," Valeti recalls. "There was a birthday and then there was a death-day—all in the same place and time." The juxtaposition of such elation and such pain planted a seed in his mind, a seed that would take years to germinate.

The grandson of an Indian freedom fighter who campaigned for independence alongside Mahatma Gandhi, Valeti grew up with a deep sense of purpose to help those in need, including animals. Part of his love for them came from his father, a veterinarian, and he also cultivated a passion for science inherited from his mother, a physics teacher.

The Valetis didn't eat beef because of their Hinduism, but they would eat chicken, lamb, fish, shrimp, and other non-holy animals, mostly on the weekends, as did many families in their neighborhood. So it wasn't until years later, when Valeti started medical school, that he began seriously deliberating his meat consumption.

"I'd read about the inefficiency of meat-eating compared to a vegetarian diet, but what bothered me more than the wastefulness was the sheer scale of suffering of the animals," Valeti remembers. "It pained me so much to see them at the markets, lined up and marched to their death, which it seemed to me they knew very well was coming."

As a result, Valeti decided to adopt the diet of Gandhi and stop eating animals altogether. He proceeded in medical school fueled by zeal to help his fellow humans, yet he also told himself that if there was one big problem he wanted to solve, beyond his career of healing the sick, it was to prevent the type of animal suffering that had so scarred him.

Throughout his medical training, Valeti struggled to remain a vegetarian, regularly tempted by both his palate and peer pressure. "I believed it was the right thing to do, but I just had trouble aligning my diet with my ethics," he recalls. "Ironically, I had to leave the most vegetarian nation on earth and come to the biggest meat-eating one before I fully stopped eating animals."

Upon finishing medical school, Valeti arrived in the United States in 1996, ready to further his career dream of training in cardiology at the Mayo Clinic. "I was mesmerized by how heart muscle would regenerate itself after I injected stem cells into my patients' hearts," he says. "It was a revelation to me. I asked, why can't we do the same thing for other muscle cells outside the body?"

So Valeti started talking with his colleagues about such a possibility. "Why couldn't we get muscles to grow in a culture?" he asked. "We could grow meat directly from meat cells into the beef, pork, and poultry that people love."

There were many protesters among his colleagues. *Are you*

serious—who would eat that? he recollects a fellow doctor asking, not trying to veil his incredulity. Even the more polite responses were generally couched in aversion to the idea. Most often people were too busy to think about it further. *I've got cardiac arrests to fix*, he was told. *We're saving people's lives here, Uma.*

Undeterred, the young physician persisted. As Valeti researched the concept more, he wondered if it would be possible not only to grow muscle tissue outside the body, but also if that meat could be tweaked to be more healthful than the meat most people eat.

"I knew that the poor diets and unhealthy fats and refined carbs that my patients were eating were killing them," he says, "but so many seemed totally unwilling to eat less or no meat. Some actually told me they'd rather live a shorter life than stop eating the meats they loved." This pushback, as well as his own struggle with avoiding meat, made Valeti fantasize about a solution that would allow people to eat the meats they wanted, but without taking such enormous health risks.

We may be taught to fear violent crime and terrorism, but the most serious threat we face is really from our own forks and knives. The number one killer of Americans is heart disease, which overwhelming evidence ties to a meat-heavy diet. Overconsumption of meat certainly isn't the only cause of heart attacks, but it's a major culprit. That's why the American Heart Association touts "the role of plant-based foods in a healthy dietary pattern" and encourages all of us to use tactics like Meatless Monday to cut our risk of having to go under the knife.

Another major killer of Americans, cancer, is also connected to our desire to gorge ourselves on meat. The World Health Organization in 2016 classified processed meats as Group 1 carcinogens, meaning we're as certain they cause cancer as we are about cigarettes. Even before that bombshell new classification, the American Institute for Cancer Research was particularly unequivocal on the topic, saying that "when it comes to American health, the research shows one thing very clearly: we all need to eat more plants and less meat."

Eating so much saturated fat and cholesterol is just a really bad idea, and likely to land you on someone like Uma Valeti's operating table. But maybe it's possible to culture meat that's just like conventional meat—with one exception.

"The main difference I thought I'd want for this meat I was envisioning was that it'd have to be leaner and more protein-packed than a cut of supermarket meat, since there's a large amount of saturated fat in that meat," says Valeti, pointing out that clean meat purveyors have their choice as to what type of fats they add to their grown muscle. (Recall that Mark Post's burger was pure muscle—no fat.) Valeti plans for something else entirely. "Why not have fats that are proven to be better for health and longevity, like omega-3s? We want to be not just like conventional meat but healthier than conventional meat."

In other words, rather than producing burgers that cause heart attacks, he imagines burgers that actually prevent them. When you're building the meat from the cell up, you have much more control over just what kind of meat you want. For example, we know that the saturated fat so often found in meat today is implicated in causing the arterial plaque buildup that leads to heart disease. Tinkering with a cow's diet might modestly change the fat makeup of the marbling in her muscle, but if you're growing her muscle on its own, there may not be much reason to add those dangerous saturated fats. Instead, you could marble the muscle tissues with healthier monounsaturated fats like the kind found in olive oil, or even omega-3 fatty acids like those found in flax seeds. In that way, clean meat purveyors could even challenge medical recommendations about what comprises a healthy diet.

Valeti's world was coming full circle. He thought that his dual dreams of sparing animals and preventing his patients from ever showing up on his operating room table might soon, very slowly, start inching closer to reality.

So one morning in 2005, still in his early thirties and after just graduating from the Mayo Clinic, Valeti was daydreaming about the potential cell-cultured meat had to do good in the world. He wondered if others shared this dream and turned to Google to find out. To his pleasant shock, he found that New Harvest had recently been formed to turn such fantasies into reality.

After devouring the New Harvest website, he shot off an email to Jason Matheny, letting the early cultured-meat prophet know about his medical background, and expressing interest in helping his mission.

The response was rapid and enthusiastic; Valeti booked a ticket to DC.

"Jason's a genius," Valeti recalls of his first meeting with the public health expert. "Here I was a trained doctor, and this guy who had never spent a day in medical school knew so much more about tissue engineering than I did."

Matheny was equally impressed with Valeti and asked him to join New Harvest's board. Once there, Valeti realized lots of people around the world shared their mutual interest in recreating the meat industry by developing the new field of cellular agriculture. With virtually no one working in earnest on such innovations, his role, he hoped, would be to motivate academics and others to jump into the pool and start putting research dollars into investigating the issue.

Numerous tissue engineers told him it was possible to do what he wanted, but in the absence of research funding or other financial support, no one was going to devote their career to this. Who knew if there'd even be a market for such a product? The "ick" factor was just too great, he was told. It was one thing to experiment on cultured tissues or even use them medically in a patient, but human consumption was just another matter.

"To put it frankly," Valeti says, "there was a lot to be made in the medical space through tissue engineering, but few people realized the same could be true in the food space."

The years went on, but the progress just seemed so minimal compared to the need. Modern Meadow began producing steak chips but then pivoted toward leather. And Mark Post at that time was primarily focused on academic research, not commercialization and certainly not marketing. No one was moving with any sense of urgency to bring animal-free meat to market, and meat was where the greatest need for disruption lay. Valeti began to reevaluate just what role he might want to play in this barely existent industry. For years he'd served as a New Harvest board member, but now he wondered if that was enough. His dream from medical school was too distant from becoming reality, and he felt like he wanted to be more than a cheerleader on the sidelines. Valeti began realizing that he wanted to be on the field himself.

After a decade of seeing so little progress toward actually putting clean meat on supermarket shelves, and already now a successful cardiologist at the University of Minnesota, Valeti knew it was time to put his money where he wanted people's mouths to be. So in 2015, Valeti opened his own dedicated cultured-meat research lab at the university, along with Nick Genovese, a poultry-farmer-turned-vegetarian-turned-stem-cell-biologist who loved the taste of meat, but just didn't want to kill animals in order to enjoy it.

Research indicated that the serious focus and scale-up efforts needed for this endeavor weren't suited to an academic environment like Post's lab. "In academia," Valeti says, "the emphasis is on grants, publications, tenure, and crippling university overheads and indirect costs." In addition, many of the systems were set up to support organ-regeneration work, not what Valeti and Genovese were trying to accelerate.

At this point, only a short year after opening his lab, Valeti realized that if he didn't move his operation out of the academic world, he might never actually replace factory farms with clean meat breweries and so was forced to make one of the most important decisions of his life: Should he put his cardiology career on hold to pursue entrepreneurship?

"I'd been thinking about this since I was twelve," he says. "I had a phenomenal career in cardiology and my most financially lucrative decades of practice were right before my eyes." The thought of foregoing the exciting innovation and lifesaving work he was pursuing as a leading cardiologist, and the substantial income he was sure to make doing so, gave him pause. In addition to his successful practice, Valeti had ascended to leadership in many national and local organizations such as the American College of Cardiology, American Heart Association, and more. He also had two small children, Neel and Tara, who had perhaps two decades of schooling ahead of them. Valeti stayed up at night contemplating the right thing to do, talking with many close friends and family about his dilemma.

One evening, he sat down at the kitchen table with his wife, Mrunalini, a pediatric eye surgeon in the Twin Cities, hoping to reach a conclusion. "Look, Uma," she said at the table, looking squarely at her husband. To hear him tell it, it was like a sentimental scene straight out of a movie. "We've been wanting to do this forever. I don't ever want us to look back on why we didn't have the courage to work on an idea that could make this world kinder and better for our children and their generation."

With her support, he was persuaded. "I recognized that there are twenty-five thousand cardiologists in the United States, yet just a handful of people working on such an important issue like ending the myriad ill effects of the meat industry. With Mrunalini backing me, I knew what I had to do."

———————

Valeti and Genovese began assembling a team to start a venture, interviewing many PhDs with deep experience in skeletal muscle biology. In the end, they brought on Will Clem, a tissue scientist who's about as far from a vegetarian as possible. In addition to his PhD in biomedical engineering, Clem is a certified barbecue pitmaster whose family owns Whitts Barbecue, a Memphis-based chain of several dozen barbecue joints.

And thus, in late 2015, the world's first company devoted exclusively to commercializing clean meat was born. (Recall that while Mark Post and Peter Verstrate debuted their burger to the world in 2013, their commercial venture, Mosa Meat, was founded in the Netherlands in mid-2016.)

With Modern Meadow now solely focused on leather, Crevi Foods essentially became America's face for this new type of meat. The Latin name—*crevi* essentially means "to arise," or "to spring forth"—appealed to Valeti's medical training. Echoing Churchill's thoughts from eighty years ago, Valeti explains the company's founding question: "In this day and age do we still need to grow a whole organism to break it down into little pieces for the simple tissue that we like to eat? Or can we fundamentally change this paradigm and grow this tissue from the building blocks of life: cells? Our goal is to produce a world where we can harvest animal products without harming life."

The next step was securing funding.

Valeti got connected to IndieBio, the accelerator program intended to support brand-new biotech start-ups and backed by SOS Ventures (now simply called SOSV), to let them know about his new start-up company idea. Ryan Bethencourt, a program director and venture partner at IndieBio, recalls the introduction.

"Most people are going to want to eat meat, no matter how tasty we can make plant-based meats," Bethencourt says. The late-thirties son of Cuban immigrants has spent his career in biotech, starting his own companies and investing in others through IndieBio, which he cofounded to solve some of humanity's most intractable problems. A vegan himself, Bethencourt is bullish on the clean-animal product market, not for his fellow vegans, but for the more than nine out of ten people who do eat animal products. "It's tough to get people to think of plant-based foods as 'meat,' no matter how good it gets," the venture capitalist observes. "People will still view it as 'fake meat.' You need meat that's biologically identical to animal meat and we can do that with biotech. It is meat; it's actual meat. It's just cleaner, safer, and truly more humane than the meat we eat today. When I was introduced to Uma, I thought that this concept and the science they were working on could remove literally billions of animals from the factory-farming system and I had to do whatever I could to help Uma and Nick on their mission, starting with getting them funded."

After less than an hour of the email introduction, Bethencourt and the IndieBio team asked Valeti to meet. Within the week, Bethencourt offered the new company its first investment check and a lab in downtown San Francisco to start making some meat. With this, Crevi Foods opened its doors and got straight to work. (Valeti had opened the office and lab in Northern California to be closer to his financial backers but was still spending part of his time with his family in Minnesota.)

At just about this same time, in October 2015, the Good Food Institute was being formed to promote plant-based and cellular alternatives to animal-based meat, dairy, and eggs. The brainchild of Mercy For Animals (MFA), the LA-based animal protection group, GFI is an independent nonprofit organization that operates like a "combined think tank and accelerator for the clean meat and plant-based meat market sectors," explains GFI executive Bruce Friedrich, who was recruited by MFA to launch the organization and serve as its first executive director.

"The clean-meat and plant-based protein companies have the potential to spare huge numbers of animals from factory farms and slaughter plants, so we wanted to do what we could to support them," Nathan Runkle, MFA's president tells me.

Almost as soon as the company was named, there were some who questioned whether it was time for a name change. Importantly, Arvind Gupta, the managing director of IndieBio, insisted that Crevi just wouldn't do. After kicking around some ideas like Minneapolis Meats and others, eventually the team starting coalescing around Memphis Meats. Not only did they like the nod to Clem's Memphis meat-loving barbecue heritage, but they also envisioned it alluding to the ancient Egyptian city of Memphis, one of the early origins of human civilization. GFI ran some online testing to see how Crevi performed against Memphis and a few other options, finding that Memphis performed the best, and the company has since been known as Memphis Meats.

Its first project involved growing beef and pork by taking the cow and pig cells that, just like in the case of Valeti's heart injections, were capable of renewing themselves. Using myosatellite cells (again the precursors to skeletal muscle cells), the researchers began experimenting.

Under the microscope, the researchers found that their cells looked identical to those from actual living animals. They had the same phenotype, meaning that they had the same features of the tissue we're accustomed to eating. "These are the same muscle tissues we eat when we eat meat," Genovese assures skeptics. The real question for him would be whether they sizzled the same way in the pan, producing the same smell and taste as conventional meat. He knew from Post's burger demonstration that this was possible, but without actually pressing the flesh to the pan, it was all still theoretical to the new company.

Upon anchoring the starter cells with small pieces of supporting materials, the Memphis Meats team started growing meat big enough to see with the naked eye. Soon they were adding physiologic steps to their cultivation techniques that allowed the meat to grow similarly to how a calf would grow into an adult cow by eating grass and other plants. They then started adding nutrients like peptides, vitamins, minerals, and sugar, along with oxygen, and began producing larger quantities of meat that they were soon regularly cooking and eating in private, trying to get the recipe just right. At this point they weren't yet experimenting with making healthier meat; they just wanted to hone their meat-making skills so they could understand how best to mimic the body's muscle-growing process before trying to improve on it.

The team continued experimenting. Just how much time was needed to get the right texture? How could they support optimal cell growth and optimize the quality of the meat? What was the right amount of oxygen to add to the culture?

Valeti and Genovese noticed that if they harvested the meat earlier in the growing process the flesh was more tender, similar to when taken from a younger animal like a calf or lamb. If they waited a little longer, it was more texturized, like that of an older animal. As Memphis Meats senior scientist Eric Schulze describes it, "The cells don't know they aren't in an animal anymore. We have to train them to realize that."

Just as novelist Isaac Bashevis Singer famously observed that "the waste basket is the writer's best friend," the Memphis Meat lab kept producing some pretty expensive scraps of cultured meat that were simply far from ready for a primetime debut. Genovese recalls that it was akin to trying to brew a new type of beer without a guidebook: you just keep brewing in different ways until you get it just right.

And they finally did get it right—or at least right enough that they were ready to come out to the world. Rather than holding a big press conference, as Post had done nearly three years earlier in London, Valeti worked with GFI to hold a private tasting event, which also included video production and a photo shoot. GFI's Friedrich offered an exclusive on the story to the *Wall Street Journal* and pitched the idea directly to the paper's agriculture reporter Jacob Bunge.

After much trial and error, and an intellectual property portfolio that was growing by the day, it was time to let the world know their big news: Memphis Meats had created the first-ever cultured meatball.

Another breakthrough came in the costs of production. While Mark Post's burger had carried a staggering price tag north of $300,000, Memphis Meats' meatball was a comparative bargain, costing the company just about $1,200 to produce. It was still far away from showing up on any Italian restaurant menus, but they were getting closer.

"In private we'd already grown, cooked, and tasted meatballs and fajitas, and that was a watershed moment," Valeti declared. "Once we put it in our mouths, it was abundantly clear that within a few seconds it had a very distinct meat flavor that I'd forgotten about. I'd been eating meat analogs, which seemed meaty to me, but this experience made me realize that they're not really exactly as meat-like as I'd thought."

In December 2015, Memphis Meats hosted its first tasting in San Francisco. "This is absolutely the future of meat," declared the company's first-ever press release, issued about a month later. "We plan to do to animal agriculture what the car did to the horse and buggy. Cultured meat will completely replace the status quo and make raising animals to eat them simply unthinkable."

The world took notice.

"Sizzling Steaks May Soon Be Lab-Grown," Bunge's *Wall Street Journal* story proclaimed. Following up on that, *Fortune* predicted, "You

Could Be Eating Lab-Grown Meat in Just Five Years." But *Newsweek*'s title perhaps put it best: "Lab-Grown Beef Will Save the Planet—and Be a Billion-Dollar Business."

The headlines even caught the attention of celebrity philosopher Sam Harris, who hosted Valeti on his popular podcast, *Waking Up*, with an episode entitled, "Meat Without Misery." On the show, Harris described his conviction that animal agriculture is an enormous cause of suffering but admitted his own difficulty in sticking to a vegetarian diet. His enthusiasm for clean meat was as strong as his prediction about the future.

"As we are scandalized by the slaveholders in our recent past, our descendants will be horrified to know what we did with factory farming," Harris declared to Valeti and his listeners. "The way we mistreated and killed billions of animals in a way that we manage to do more or less with a clear conscience simply because we were keeping the details out of sight and out of mind . . . It doesn't surprise me at all that there is or will be a huge market for this if you can accomplish your aims."

Harris's fellow rationalist-in-arms Richard Dawkins has expressed similar sentiments. In 2013, in response to a question I put to him on meat-eating, Dawkins was pretty straightforward. "I would like everybody to be a vegetarian," the evolutionary biologist asserted. "In one hundred or two hundred years' time, we may look back on the way we treated animals today as something like we today look back on the way our forefathers treated slaves."

The fact that such rigorous thinkers as Harris and Dawkins believe that our present-day meat-eating will one day be viewed similarly to the slave system of our ancestors and yet still struggle to become vegetarians themselves—while conceding it's the right thing to do—affirms the argument of many in the clean meat space. When even those who are persuaded that they shouldn't eat animal-produced meat for some reason feel compelled to continue eating it, foods like Uma Valeti's may help many bridge the gap between their values and their actions.

———————

Producing a meatball and revealing the first news of the tasting to the *Wall Street Journal* was one thing. It's another beast entirely to go from prototype to grocery store shelves. Memphis Meats now had to secure additional funding to even start thinking about turning its publicity stunt into a marketing reality. Though the company had aimed to raise $2 million, Valeti notes, "We surpassed even our own expectations." With meatballs now in hand, they quickly raised more than $3 million—from investors like SOSV, Jeremy Coller, and New Crop Capital, a $25 million venture capital fund run by GFI's Friedrich and his board chair Nick Cooney.

New Crop's investment of $500,000 into the infant company was particularly noteworthy, as one of the fund's biggest investments to date. "We see Memphis Meats as the future of meat," Friedrich tells me. "By getting in on the ground floor, investors are positioning themselves for a big return, and they're also enabling a potential solution to two of the world's most pressing problems: feeding almost ten billion people by 2050 and helping the nations of the world meet their climate change obligations under the Paris Agreement."

Perhaps even more interesting was the fact that Jeremy Coller, the hugely influential British finance executive, was putting his own wealth into the company. With his early investment in Memphis Meats, Coller sent a signal heard by many in the investment world. "When Coller speaks, investors listen," says Chuck Laue, the co-founder and vice chairman of Asurion, one of the biggest insurance operations on the planet. "The fact that Coller is so concerned about factory farming and its effects on the planet, and that he's so bullish about these cultured-animal product companies, attracts a lot of other investors to the space."

Laue and his wife, Jennifer, have their own venture fund, Stray Dog Capital, and additionally became early investors in Memphis Meats. (He also serves on the Humane Society of the United States' board of directors, and she on HSUS's National Council.) In fact, Stray Dog Capital only invests in start-ups it sees as advancing animals' interests, with the bulk of its millions of dollars of investments being in food companies—both plant-based and cellular ag—seeking to compete with animal agriculture.

For several years, Coller—whose private equity firm has assets under management of $17 billion—has railed against the environmental harm done by the animal-agriculture sector. After investing in Hampton Creek following a 2014 meeting with Josh Balk, Coller began meeting more and more influentials in both the plant-based and cultured-meat worlds, and continued to back those he found most promising.

"You need solutions if you're going to end animal factory farming," the private equity titan proclaims. "These companies are those solutions, and many of them are likely to do very well."

Ironically the son of a leather coat-maker, Coller, a vegetarian, now spends considerable resources to raise awareness about the unsustainable nature of animal agribusiness. A friend even wrote two obituaries for him: one in which he died at fifty-four and another at ninety-eight. The description of him in the first one made him out to be "a total bore" in Coller's words, while the second one named him as a key reason the factory farming of animals came to an end. Increasingly, Coller's charitable foundation and personal investments are now aimed toward the same goal. As he explains:

Factory farming is a major risk for investors. There are four inconvenient truths about factory farming—what I call the four horsemen of the apocalypse: human health, climate change, food security, and planetary resources. Factory-farmed meat is the number one user of freshwater, antibiotics, deforestation, and it just isn't an efficient way to feed people. Farm animals are outpacing humans in demand for cereals now—we need to stop this madness.

Just a year after the big meatball unveiling, in March 2017, Valeti announced to the world, again through a Jacob Bunge *Wall Street Journal* exclusive, that he'd now produced and served the first clean Southern fried chicken and duck à l'orange. No longer were the companies focused just on beef: it seemed that poultry was soon to get its due as well. Taste testers came away from the unveiling satisfied, saying they'd definitely eat it again. And the best part: on a per pound basis, Valeti had already halved his costs from the meatball unveiling a year earlier. Whereas Mark Post's hamburger meat initially cost $2.3 million per kilogram and Valeti's 2016 meatball was $40,000 per kilogram, this 2017 poultry meat was produced for a comparatively bargain price of just $19,000 per kilogram. (Of course, this is still exponentially more than conventional poultry meat, but Valeti's movement is rapidly heading in the right direction.)

The news coverage of Valeti's efforts continued, which in turn brought some very welcome attention from the meat industry. As noted in chapter 3, Cargill—America's largest privately held company—became the first conventional meat company ever to invest in a clean meat start-up, raising eyebrows across the protein sector. It joined billionaire investors like Bill Gates, Richard Branson, along with Jack and Suzy Welch, in a $17 million funding round for Memphis Meats, putting the company much closer to commercialization.

In a Fox Business interview, Cargill CEO David MacLennan discussed his new investment, touting that Memphis Meats "produces chicken or duck in a way that doesn't use the resources that traditional meat uses. So it's all about sustainability. Call it 'clean meat' if you will. It's a way to produce meat in a different alternative that isn't as resource-intensive."

The new investment from the meat industry itself even seemed to change the tune of former cultured-meat foe, the Animal Agriculture Alliance. Recall that a spokesperson for the Alliance called Mark Post's beef a "Frankenburger" in 2013, but after Cargill's 2017 Memphis Meats investment, the organization's CEO struck a different tone in an interview with Food Safety News. While bristling at the term "clean meat," she conceded that "given the forecasts that food production will need to double by 2050, lab-grown meat is simply an additional way to help meet that demand."

As a result of Cargill's and several others' investments, Memphis Meats set its sights with a laser-like focus on producing clean meat that it can start selling to consumers. Just like Post, Memphis Meats aims to start with ground meat items like hot dogs, hamburgers, chicken nuggets, and sausages, since the technology to produce more complex items, like a steak, hasn't yet been developed. They intend to be on the market with limited offerings in the near-term, contingent on continued funding and scaling up their work. Valeti doesn't provide a specific time line for market entry, but says they'll be making announcements on this subject as they approach commercialization. Whether that'll be in the form of distinct products advertised as clean, or cultured components blended into conventional meats to help ease the transition, Valeti still isn't sure.

What is sure, though, is that much needs to happen between now and then, including technological breakthroughs that no one knows yet how to achieve. Perhaps the biggest challenge Valeti and his crew face, however, isn't the technology. They're confident that they'll figure out how to bring the costs down even further and mass produce. The bigger challenge for them is simply how to persuade consumers that this is a food they'd actually want to eat.

How Memphis Meats will get its foods to the market in the first place is still an open question, but perhaps the more important question is whether anyone will even eat them. After all, many people instinctively recoil upon hearing about meat grown outside of animals. And as the saying goes, you don't get a second chance to make a first impression. It's very hard for people to shake an initial reaction to anything—a person, an idea, or, in this case, a new food—and that's a big obstacle for clean meat.

Some early polls gave clean-meat enthusiasts reason to be concerned. A 2014 poll conducted by the Pew Charitable Trusts upon first glance doesn't bode that well for companies like Memphis Meats. The think tank found that only 20 percent of Americans are willing to eat "meat that was grown in a lab." College-aged respondents were the most likely to give it a try, but still, less than a third expressed interest. Men were more interested in sampling the hypothetical product than women, though, like college students, two-thirds of them were still unwilling. A study of Belgian consumers, also published in 2014, found a quarter were willing to try it, with the number rising to 43 percent when informed of the environmental benefits, and another 51 percent said they then might try it. In other words, the more people learned about why clean meat is better for the planet, the more willing they became to try it, making it clear just how critical the messaging will be for Memphis Meats as it gets closer to commercialization.

But if large numbers of consumers aren't clamoring for this product, is it even apparent that there'll be a market for it? No one of course has a crystal ball, but the Pew survey may not, in fact, be an accurate assessment of what consumers will or won't want. Pew simply asked people, "Would you eat meat grown in a lab?" without explaining any of the benefits and without discussing price. The Belgian study at least showed that people can be easily swayed when given more information, while the Pew researchers made the whole thing sound like a crazy dare.

Second, as GFI's Bruce Friedrich points out in a blog on the topic, clean meat at scale won't happen in a laboratory—all processed food started in a food lab, even Corn Flakes and peanut butter, for example. But no one asks, "Would you eat lab-produced Corn Flakes?" Rather than being produced in a lab, clean meat would be produced in a factory (or call it a brewery if you prefer), where the majority of food sold in supermarkets is produced. Food companies of course have R-and-D teams laboring away in labs, but once they get their recipe down, the actual food production moves to a factory. Similarly, clean-meat factories will be a far cry from a laboratory; they'll have massive tanks in which the meat will be cultured on a huge scale.

So the question posed by these studies may be the wrong one. Really, the fact that one in five people would eat "meat that was grown in a lab"—apparently for no reason at all besides sheer curiosity—could be considered highly encouraging for the clean-meat industry. Once Uma Valeti has a chance to tell consumers about the benefits, and especially if the product is cheaper than meat from slaughtered animals, there doesn't seem to be any valid reason why consumers won't switch. Not everyone will convert, needless to say, but enough will likely do so to make a difference, and, presumably, a profit. As well, even if only twenty percent of meat-eaters were willing to switch, that would still make clean meat a multibillion dollar industry.

Despite Valeti's enthusiasm and the willingness of a lot of consumers to give clean meat a try, many well-known voices in the healthy and sustainable food world aren't on the cultured bandwagon. For some, the biggest problem in our food system is that there's too much technology involved in it already. They want to see what they consider a more natural diet take over American food habits and envision a return to an era when our food was less processed, more local, preferably organic and produced on small-scale farms, and certainly free from biotechnology.

In a 2011 interview with *Harvest Public Media*, Marion Nestle, a nutrition professor at New York University and the author of popular books on healthy eating, called the idea of eating meat grown outside an animal "revolting" in response to learning of Modern Meadow's early efforts. Nestle is a long-time advocate of plant-based eating who believes that the healthiest diet is what she calls "largely vegetarian." As an occasional meat-eater herself, she has no desire to give cultured animal products a try. "What's wrong with food?" she asked in the interview. "Food seems just fine to me. Maybe it's what the world's coming to, but I don't want it."

Nestle's revulsion to the idea of eating meat grown without animals is an instinctive one for many. The intuitive appeal of eating food that was "close to nature" has a real allure, and it's understandable to blame seemingly "unnatural" diets on the health woes that plague us today. That said, nearly none of what we eat today is truly that "natural," nor would we want it to be. For example, how many of us prefer buying watermelons with seeds compared to those conveniently bred to be unnaturally seedless? Or how many Americans would be willing to give up apples, since apple trees aren't native to North America, or would even prefer to eat the tiny, mealy apples from which today's massive and sweet fruits are quite unnaturally descended? There's a popular bumper sticker that reads, "No Farms, No Food." Well, food historian and author James McWilliams suggests that, considering how much science is involved in nearly all food production today, perhaps such drivers should also sport stickers reading, "No Scientists, No Food."

Michael Hansen, a scientist with the Consumers Union (CU), agrees with Nestle's disgust at the thought of eating Valeti's meat. He knows better than most just how serious food safety problems can be for Americans. The CU was among the leading organizations that successfully fought for the Food Safety Modernization Act, signed by President Obama in 2011, which overhauled the way America's food industry prevents food-borne contamination. And the CU is a prominent decrier of the animal-ag industry's overuse of antibiotics, a contributing factor to the problem of antibiotic resistance.

Still, in spite of the food-safety benefits and reduction of antibiotic use inherent in cultured-animal products that people like Uma Valeti tout, Hansen isn't a fan. "The market isn't going to want this, especially not millennials," he told me. "The trend is toward real food, not this synthetic stuff they're producing in a lab." Hansen asserts that the entire cultured-animal product movement is "nonsensical," and that attention to it is more about sci-fi fantasy than the reality of our current food system. "I'll believe it when I see it," the scientist proclaims. "They've been making predictions about this stuff for decades, and it never comes to fruition."

Hansen is right that predictions have been made for years about cultured meat coming to fruition, and yet the meat industry largely hasn't felt that threatened. But things do seem to be changing in the wake of high-profile product unveilings by the likes of Post and Valeti, and certainly the investment from Cargill. Gone are the days of clean meat being purely a theoretical daydream of environmentalists who want a more sustainable way to produce meat. With commercialization looking increasingly likely, we won't need to rely on pollsters to tell us how consumers may react when clean meat is available to them. People like Hansen and Nestle may not want to eat meat if it didn't come from a slaughtered animal, but how many others will share their repugnance at such a thought?

Kristopher Gasteratos, founder of the Cellular Agriculture Society (created in 2016), is more optimistic. He believes animal agriculture is so inefficient that humanity will be forced to abandon it, at least for the bulk of our protein production, or we'll pay the price. His analysis of the situation doesn't pull any punches: "Factory farming of animals will end one way or the other. The real question is this: If we don't find an alternative to factory farming soon, will we as a civilization end with it?"

Gasteratos is convinced that the public will come to accept clean meat because there's such an existential necessity for it. But his view is also informed by a study he conducted over the course of 2016 with the assistance of both New Harvest and the Good Food Institute. In the study, Gasteratos led a team of researchers who asked thousands of survey respondents their views on the topic. Based at Florida Atlantic University, the project ultimately surveyed more than thirty-two hundred undergraduate students and about fifteen hundred adults both in the United States and Australia (the two nations with the highest rates of meat consumption on a per capita basis). Unlike the aforementioned surveys, which largely asked if people would eat "meat grown in a lab," Gasteratos took a deeper dive, wording his key question in a way that provided respondents with more context:

> Scientists are working towards producing meat by using animal cells instead of living animals. This new method of harvesting meat is called "cultured meat" and will likely be available to the public within the next decade. It is important to note that cultured meat is real animal meat, so it should not be confused with current meat substitutes which are made from plants. If cultured meat is proven safe by long-term research, tastes the same as current/conventional meat and is priced affordably, would you eat cultured meat?

Upon simply being asked this question, without any discussion of clean meat's benefits, 61 percent of the university students claimed they'd either "probably" or "definitely" eat it. After being told some of the benefits, either ethical, health, or ecological, that number spiked to 77 percent. Among the fifteen hundred adults, the numbers were similar: 62 percent were willing to eat it without knowing its benefits, while 72 percent were willing once they knew of those benefits.

Other interesting findings from Gasteratos's work include some pretty fascinating results about just who is most interested in eating this meat. "People still seem to be generally unaware of this topic, but what really shocked me was our finding about how higher self-reported meat consumption correlated with higher cultured meat acceptance." Basically, the people who say they eat the most conventional meat tend to be the most receptive toward a cultured alternative, while people who say they eat little meat, and especially vegetarians and vegans, are the least interested.

In other words, clean meat probably isn't for the people shopping at the farmers' market or their local co-op. It holds far less appeal with the natural-foods crowd than the crowd going to KFC. But that's okay. In fact, it may even be for the best considering that the number of people who eat conventional meat is far, far larger than those who frequent their local farmers' markets.

Comments left by respondents offered some good qualitative insights into the general perception. "I don't care where the meat came from so long as it's safe and tastes right," explained one respondent, echoing a widely held sentiment among participants. Others expressed some qualms about meat-eating but thought cultured meat could be the answer to their concerns: "I heard meat is really bad for global warming," one respondent wrote. "This would sort of absolve me of that guilt."

And in 2017, a new survey was conducted of American adults and published in the prestigious *PLoS One* journal, with some of the most promising findings yet. Even when using the term "in vitro meat" with respondents, two-thirds said they'd probably or definitely be willing to try it. A third said they'd be "willing to eat IVM as a replacement for farmed meat" and half said they'd be "willing . . . to eat IVM compared to soy substitutes." And very interestingly, respondents largely agreed that "in vitro meat" is unnatural, but they said they thought the benefits were high, especially since it "has the potential to solve world famine problems and reduce the impact of global warming associated with farming."

One consistent aspect of all these consumer acceptance surveys is that men show higher willingness (and in some cases eagerness) to eat clean meat than women. This comports with research on acceptance of other food technologies, including genetically modified foods. Chris Bryant, a PhD candidate at University of Bath whose work focuses on public attitudes toward cultured meat, has mixed reactions to this gender divide. "This could be seen as a good thing, since men typically eat about 50 percent more meat than women," he tells me. "Although there's also evidence that women make the majority of food decisions on behalf of the family, so persuading women that the product is safe is likely to be a key goal for consumer acceptance."

Whether the Consumers Union's Hansen is right that no one will want clean meat or the researchers are right that a sizable portion of people are open-minded about it, at least some leaders in the sustainable food movement argue that no one should be eager to consume these new foods. Dana Perls, senior food and technology campaigner for the nonprofit Friends of the Earth, joins people like Nestle and Hansen in expressing real concern about cultured animal products. Perls is primarily worried about the use of biotechnology in food and agriculture, especially synthetic biology, which she refers to as "an extreme version of genetic engineering." For her, synthetic biology versions of products like vanilla, stevia, and saffron, all already on the market, represent a roll of the dice for consumers. And the fact that such products are often marketed with terms of sustainability is an additional affront to GMO critics like Perls.

She's right that genetic engineering has become prevalent in our food system, sometimes without much of a societal discussion beforehand. But there's a larger point that might be getting missed.

For one thing, it's important to note that the clean meat companies at least so far aren't using genetic engineering to produce their protein (though as we'll see in chapter 7, the same isn't necessarily so for egg- and milk-makers). But perhaps more important, while committed natural-foods advocates rail against the unnatural processes used in our agricultural system, most consumers don't actually seem to care all that much. GFI's Friedrich points out that even if the most dire predictions are right—only 20 percent acceptance of clean meat—that is still a colossal market. "Recall that plant-based meat is just one-quarter of one percent of the meat sector now. Twenty percent is forty billion dollars per year." There may be a segment of the population that wants meat from well-treated animals without any of the drugs and cruelty that plague most animal farming today, and as noted earlier, clean meat perhaps isn't for them. But, Friedrich goes on, "the vast majority of the population is eating feces-tainted meat from animals who were pumped full of drugs and treated in ways that would be considered criminal animal cruelty if committed against dogs or cats. Clean meat is the alternative to that meat."

And many sustainable food advocates seem to agree. One of industrial agriculture's biggest foes is Bob Martin, director of the Johns Hopkins Bloomberg School of Public Health's Center for a Livable Future. Martin served as the executive director of the blue-ribbon Pew Commission on Industrial Farm Animal Production, which largely reached conclusions about factory farms that advocates like Perls wholeheartedly embrace. Animal agriculture as practiced today, Martin argues, is simply unsustainable. That's why he's such an enthusiast for clean meat. "Cellular food animal production is very promising and could solve the problems caused by the present concentrated animal feeding operation model," Martin celebrates. Of course, that doesn't mean it will be an easy sell.

There's a certain "technophobia" that many people experience when they hear about new technologies they may not understand.

Some comedy fans may remember when Penn and Teller led a campaign against dihydrogen monoxide (DHMO) in an episode of their show *Penn & Teller: Bullshit!* Informing citizens that DHMO can be fatal if inhaled, may cause severe burns, is a fire retardant, and more, they had little problem persuading people to sign a petition to ban DHMO. If you know chemistry, though, you know that dihydrogen monoxide is simply two hydrogen atoms and one oxygen atom, or H_2O—the chemical name for water.

This may not be a perfect comparison to what's going on in cellular agriculture, but it does illustrate a relevant point. People often fear what they don't understand, especially when science is involved. In some ways we're eager to embrace new scientific breakthroughs, for example medical advancements and even new iterations of the latest smartphone technology. But when it comes to food, some of us seem to have an instinctually different view of technology.

New Harvest's Isha Datar puts it well: "Look at a Sam Adams brewery. The people running it are microbiologists, but you don't consider them mad scientists. They're just brewing a product we're already accustomed to consuming. And that's eventually what cultured meat purveyors will be doing, too."

In other words, a lot of foods we eat and love are produced by scientists in a factory, and we don't think twice about it. When was the last time you felt uneasy about downing an energy bar, for example? Even though that bar was likely developed by people in white lab coats and produced en masse in a factory that combined a lot of ingredients, few people pause when picking them up at the gas station. Even though some people want bars with short ingredient lists with no words that sound scientific, most consumers just want to refuel their bodies with a product that tastes good, is affordable, is safe to eat, and satiates their hunger.

———————

Consumer acceptance is certainly on Uma Valeti's mind, though simply creating a product that consumers can actually have a chance to buy is the primary task on his mind. But he doesn't want to produce just any meat. Valeti's feverishly trying to figure out a way to produce meat that is better for human health than conventionally produced meat. Adding healthy fats is a good start, but the key is to get the right fats in the right proportion so as not to negatively affect the taste and mouthfeel of the meat. But there's one way his meat is already superior to conventionally produced meat: it's just a lot cleaner.

"This may be the cleanest meat the world's ever seen," the doctor declares. He proudly shows an image of two petri dishes, both segmented into three sections. In one, each section has meat samples swabbed from conventional pork, free-range organic pork, and Memphis Meats pork. In the other, the same, but with beef. And in both cases, the results of the study are unambiguous. On the conventional as well as on the free-range organic samples, there's voluminous bacterial growth. But on the Memphis Meats portion of the two petri dishes, there's no fuzzy bacterial growth, just a pristinely clean segment.

He's done the same with whole pieces of meat too, leaving samples of conventional chicken, organic chicken, and Memphis Meats chicken on a countertop at room temperature for several days. The results were the same: ample bacterial growth on both meat samples from slaughtered birds, and virtually none on the cultured version. These experiments are leading many to wonder just how much longer clean meat's shelf life will be, as it seems to spoil at such a slower rate than conventional meat.

"It's a food safety advocate's dream come true," Valeti proclaims. "Just think of every time there's an outbreak of food-borne illness as a result of contamination from animal feces—this type of technology could prevent that tragedy."

The potential for that fantasy to be realized could largely depend on what regulatory hurdles, if any, are erected to slow the movement toward cultured meat. Would the FDA even approve the sale of cultured meat? Or would the regulating body be USDA instead? Are there state regulators who'd need to be involved as well? Would there be some type of movement to derail it, just as there is with genetically modified organisms (GMOs) today?

Valeti is quick to point out that culturing meat doesn't require GMO technology, but in a world where misperceptions can slow down innovations, he's taking nothing for granted. He says he's intent on "working with everyone who truly would look at the technology we're developing with the lens of the greatest benefit for humanity and our planet."

Despite the fact that his meat doesn't use GMO technology, the debate over genetically modified animals might still illustrate potential potholes in Valeti's road to commercialization. Take, for example, the AquAdvantage salmon. Genetically modified to grow to full size in sixteen to eighteen months rather than the three years it typically takes, the fish was approved for commercial sale by the FDA in 2015 and in Canada in 2016. Animal welfare advocates decried the decision since the fish can suffer many problems from their rapid growth, but supporters of the transgenic process cheered. The fish was first submitted for FDA approval twenty years prior to its approval, despite the FDA concluding the fish is safe for humans to eat. If that's how long it took to approve raising a fish that's been selected to grow faster, what problems might a company like Memphis Meats have when it's producing animal products without any animals at all?

Bruce Friedrich, who worked against the approval of the GMO salmon when he was working for an animal welfare organization prior to GFI, isn't concerned. "That was an entirely different situation," he explains. "In that case, FDA used the new-drug approval process for a variety of reasons, including the fact that the company was actually slicing many genes from various animals together to create an entirely different animal. There were all kinds of animal welfare and environmental questions that won't apply to clean meat, and there were health questions that apply when you're creating a new product—a fish who is very different from any other fish in nature—but that won't apply to the creating of meat that is identical to other meat."

At this point, we don't even yet know if clean meat will be regulated by the USDA or FDA; the former regulates meat, but its regulatory framework assumes that there are live animals involved and slaughter processes to inspect. The latter regulates all other food, including fish. Anticipating the problem is part of the reason Memphis Meats hired Eric Schulze, who'd previously spent six years at the FDA evaluating novel biotech for the agency. "I'm confident there is a regulatory pathway for the company to go forward," he says.

One reason for the confidence is because the food safety benefits the product will bring could make it just too appealing for regulators to keep such products off the market. If meat that's riddled with fecal contamination is routinely allowed to be sold to consumers, why would meat free from such adulteration be excluded? The problem of feces in meat is so pervasive that the industry knows it can't prevent it. Instead, it lobbied for years to be allowed just to irradiate its meat before it reaches the consumer (in other words, to apply ionizing radiation to the meat to kill bacteria). As meat sector spokespeople correctly point out, when you have live animals, you're going to have feces and other bacterial loads present, and that can get on the meat. Divorcing meat growth from the live animal could eliminate this problem altogether.

Another reason for optimism is that the government already permits similar technologies as safe for use in human medicine and other products. As is the case with rennet in cheese (to be discussed more in chapter 7), such biotech processes are already approved for food that's routinely consumed. The biotech ship has sailed. Whether it's in our medicine or our food, we're all using and benefiting from scientific advances in our daily lives. Given that culturing technology allows us to produce meat that is essentially identical to what we're used to, it's not too difficult to see how these products could win regulatory approval.

Even if you see a pathway for jumping regulatory hurdles, there's still the technological hurdles that Memphis Meats needs to clear. As discussed earlier, moving away from fetal bovine serum is one such critical advancement, and one that Memphis Meats is making great progress on. In February 2017, Genovese published an article in the journal *Scientific Reports* (part of the Nature publishing group) on how to grow muscle tissue without animal serum at all, using a synthesized version of the serum that contains no animal ingredients. "Our goal is to entirely remove the animal from the meat production process," says David Kay, head of mission and business analyst at Memphis Meats. In other words, even though the company's early research involved serum to help facilitate cell growth, Memphis Meats knows that without replacing that serum with a cheaper animal-free nutrient media, they won't be able to successfully commercialize and compete with conventional meat.

"Animal-free (both serum-free and animal-product-free) growth media are already on the market for many cell lines," the Good Food Institute's senior scientist Christie Lagally tells me. "This is an example of how we, as humans, have already figured out how to make animal-free media."

But cell needs are just one of the issues these entrepreneurs face, especially the meat-makers. "Turning collagen into leather is hard," Modern Meadow's Andras Forgacs explains. "It's not like you just throw some collagen in a petri dish, come back and voila, you've got a leather coat. But I'll tell you this: it's a lot easier than trying to tissue engineer meat."

With proofs of concept now in hand—and in stomach—from

start-ups like Memphis Meats and Mosa Meat, no one still suggests that it's technologically impossible to produce animal products in the lab—even if it's not easy and we're still a long way off from creating a chicken breast or steak. But the question on which the success of cultured meat depends is whether it's possible to produce it at anywhere near the price point of the animal products with which they will compete.

Currently, commercial space flight is technologically possible, but only for the wealthiest cosmic tourists. Similarly here, if Warren Buffet wants to buy some clean meat, Uma Valeti is ready to deliver. But in order to make his meat available to the masses, Valeti will need to bring his costs down tremendously. Valeti feels confident on this point, observing that the first human genome cost $3 billion to complete and only fifteen years later you and I can each get our personal genome fully mapped for a few thousand dollars, and if we're willing to settle for just the basics about our genes, it'll only put us out a couple hundred bucks plus a tube's worth of saliva.

"The first iPhone cost $2.6 billion in R-and-D—a lot more than the first cultured burger," says Bruce Friedrich. "The early stages of all technology cost a fortune."

Friedrich's point is well taken. In fact, the iPhone is a good example. The technologies in it are millions of times more powerful than the first generations of computers and yet are hundreds of times less expensive to produce today. Similarly, some estimates of Instagram's editing features alone suggest that they'd have cost a consumer $2 million a decade ago. Now, it's just a free download. Already Mark Post's $330,000 burger is nearly 80 percent cheaper than its progenitor. Just how he's done it is a proprietary matter for Mosa Meat, and there's still a long way to compete with commodity meat, but given the trajectory he's on, he's optimistic.

In fairness, others are more pessimistic about ever reaching reasonable price points for cultured meat. One of the skeptics is synthetic biologist Christina Agapakis. The scientist, who earned her PhD in biomedical sciences at Harvard, wrote a piece denouncing what she sees as the false promise of cultured meat. "Cell culture is one of the most expensive and resource-intensive techniques in modern biology," she warns. "Keeping the cells warm, healthy, well-fed, and free of contamination takes incredible labor and energy." It's one thing to spend like that for medically therapeutic uses, but no one's going to drop those kind of dollars for food. "Grand technological fixes can look good if you don't peer too close at their workings. But as should be clear once you examine the case of in vitro meat, the meat problem won't really be solved with flashy tech." The only real solution, Agapakis believes, is for humans just to eat less meat.

She's not alone. GiveWell, the charity evaluator nonprofit that helps people make cost-effective choices when donating to charity, published its own analysis of the prospects for cultured meat to replace the factory-farming industry, which it believes should be displaced. Its conclusion was brighter for cultured eggs and milk (which we'll discuss in chapter 7), but when it comes to the meat companies, "we currently see developing cost-competitive cultured muscle tissue products as extremely challenging, and we have been unable to find any concrete paths forward that seem likely to achieve that goal."

Perhaps the proof is in the pudding. Although GiveWell found in 2015 that the science wasn't ready, two years later, many leading tech investors were convinced that clean meat could become price competitive with conventional animal meat. Perhaps part of their reasoning was due to Dr. Liz Specht, the Good Food Institute's senior scientist, who met with more than twenty-five venture capital firms and investors to share her analysis of the economic viability of clean meat, including most of Memphis Meats' soon-to-be investors.

The entire clean meat community knows that price points need to come down, and they all but dismiss those who suggest that they can't, pointing out that considered attention to cultured meat is very new. After all, excluding Modern Meadow's brief and now abandoned experiment with steak chips, the very first clean meat company was formed in late 2015, and more than half of the money that's ever been spent in this field has come in the past few years.

And why shouldn't the price come down? The inherent efficiency advantages that growing animal meat has over raising whole animals puts companies like Memphis Meats in an enviable position. They don't yet have economies of scale nor the knowledge of how they're going to get there, but it takes so much more land, water, oil, and more to raise animals that it's not difficult to see these companies eventually competing on cost. What's needed now is the R-and-D resources to figure out how to do it.

In addition to the fact that the technology they use is so new, companies like Memphis Meats are also unlikely to be the beneficiaries of subsidies from the federal government in the way conventional meat producers are. Most of that assistance comes via the farm bill, legislation passed every five or so years that sets ag policy for the nation. As policy expert Lewis Bollard of the Open Philanthropy Project observes, "The farm bill's centerpiece is farm subsidies, which mainly subsidize farmers to buy crop insurance. These insurance subsidies lower the price of crops like corn, helping factory farmers, for whom feed costs can reach 70 percent of production costs. Farm subsidies cost the American taxpayer about $20 billion every year, more than double the EPA's budget, mostly to support wealthy corporate farms." But, Bollard continues, $20 billion may sound large, yet it may not actually do as much as one would imagine to lower meat prices.

When agricultural subsidies began in the 1930s, then secretary of agriculture Henry Wallace called them "a temporary solution to deal with an emergency." That emergency was the Great Depression and the Dust Bowl, which really did threaten the American food production system. Yet that temporary solution has now lasted nearly a century, in both good and bad times for farmers, and even today, when the average farm household incomes greatly exceed the national average.

The big ingredient fed to farm animals, corn, is perhaps the king of the subsidy game, yielding billions in federal handouts. To be clear, only a tiny percentage of corn grown in the United States ends up on grills at backyard barbecues. Just like with soybeans, a large portion of corn we grow ends up in the stomachs of the chickens, turkeys, pigs, and cattle who wind up on those barbecue grills instead.

In other words, the meat industry receives indirect agricultural subsidies that at least somewhat artificially reduce the cost of the most expensive part of their business: the corn and soy grown to feed billions of animals. But just how much ending such livestock feed subsidies would increase the cost of meat is unclear. Some agricultural economists, like Purdue University's Jayson Lusk, contend it would perhaps only increase prices by 1 percent. But there are other subsidies that aid industrial animal ag, too, such as "surplus buy-ups" in which USDA helps industries that produce more than consumers want by purchasing their surplus eggs, pork, or other unwanted commodities. The foods then get dumped into federal prison cafeteria programs and other federal food programs. And perhaps more importantly, many of the costs of animal agriculture, including environmental and public health, are largely externalized and not reflected in meat prices.

Whether or not agriculture subsidies have a significant impact on meat prices, they don't seem likely to go away anytime soon. Both Presidents George W. Bush and Barack Obama favored substantially reforming ag subsidies and were backed by the editorial boards of the most influential newspapers, like the *Washington Post* and *New York Times*. It may be common sense that such handouts are market distorting and unfair, but the farm bill is largely written by congressional agriculture committees. Those committees are often dominated by rural lawmakers who are beholden to agribusiness interests that both comprise much of the business in their districts and are major political campaign contributors. Perhaps that will change, but as of now, it's hard to envision near-term farm bills not offering the type of lavish handouts that have been given to agribusinesses for nearly a century.

Subsidies or not though, the efficiencies cellular ag brings may end up so overwhelming, they could just compensate even without benefiting from the farm bill in the way conventional meat producers do. This will be particularly true if the meat giants get in the cellular-ag game themselves and therefore have an interest in helping them gain market share.

"Think about it," Valeti argues, "meat producers are just protein producers. Even if they don't care about the ethical or environmental benefits our proteins bring, I bet the companies would probably love to produce meat so much more efficiently while also making it cleaner, safer, and better."

Valeti predicts that there'll be a price premium at first for clean meat, but that such premiums will soon vanish. "There's just so much inefficiency in meat production," he says. "It takes twenty-three calories of inputs to produce one calorie of beef. Our production techniques are aiming to make this three to one."

In the end, there's no getting around that animal protein is incredibly resource-intensive to produce. But when we're able to grow just the parts of the animals we really want—their meat, for example—and don't have to worry about producing the skeleton, brain, intestines, and other less-desirable parts, we no longer require nearly as many resources.

Of course it really boils down to whether consumers will actually eat Memphis Meats' offerings. On this point, Valeti is confident. "Right now people eat meat with their eyes squeezed shut," he says.

Disputing the 2014 Pew polling results, he presses back in defense of consumer acceptance. "They don't think about the inefficiency, the filth, cruelty, the climate change. But once they know there's an alternative that's healthier, that doesn't include the pathogens, and that doesn't harm animals, people will absolutely switch over."

There's already one restaurant chain eager to sell Memphis Meats' products: Clem's family's Tennessee barbecue chain. Clem has the vision: "The animal could stay there and be the spokesperson, the mascot. You could point to him and say, 'That's where your meat came from.' We could have a pig mascot, and he'd still be alive," laughs Clem, unknowingly copying Mark Post's original idea to bring a live pig to his press conference and serve sausage made from his cells.

There are innumerable challenges facing Memphis Meats and companies like it. From technological and regulatory barriers to cost and consumer acceptance hurdles, Valeti knows that success is hardly assured. But reflecting on how far he's come since that fateful birthday/death-day that so moved him three decades ago, he now has high hopes for the future. "Within twenty years," he projects, "a majority of the meat eaten in America will be slaughter-free."

Clem, too, has a prediction for what such a world would look like. "A lot of restaurants now have a beer tank in the corner, and they're brewing an IPA," he says. "Well, this will be the same thing, only the tank won't be brewing beer; it'll be growing beef, pork, or chicken instead."

Motivated by the support for their work and flush with funding, Valeti and Genovese work closely with their growing team in their Bay Area headquarters with quiet vessels of various animal meats being brewed in their state-of-the-art mini-meat production facility. In August 2017, as I sat in their office looking at walls adorned with blown-up canvas portraits of various Memphis Meats products, an apron-clad employee emerged from the kitchen. "Want to try some clean duck?" he asked with a smile.

My host offered a plate with two pieces of warm duck meat, recently grown and cooked just yards away from where we were seated. To the side of the plate was a cilantro chimichurri dipping sauce.

"Do you really think I'd obscure the taste with a sauce?" I joked. "How much did this cost to produce, anyway?"

Valeti hadn't done the math specifically on these two pieces of meat, but he assured me that it was less than I might think.

Slowly piercing the first piece with my fork, I watched it "sweat" the way conventional meat does when cooked. As I pressed it against the roof of my mouth with my tongue, a deeply savory and meaty sensation arose, which chewing quickly enhanced. It was exactly how I remembered duck tasting from my youth. The second piece I tore apart with my fingers to see its insides, noticing the stringiness that held the meat together, just like a regular piece of meat. After some photo-taking, it, too, made its way into my digestive tract. Valeti smiled as he videotaped the experience with his phone.

He remarked on the distance he's traveled from his days as a student in India, visiting the live animal markets that haunted his memories for years. Valeti's dream of sparing patients from the operating knife and animals from the slaughter knife is inching closer every day. On this particular afternoon, the evidence of that dream's impending actualization took the form of two duck nuggets, which I thoroughly enjoyed.

6

PROJECT JAKE

There are clearly plenty of questions about whether bringing clean meat to market is feasible. Can companies like Memphis Meats bring their costs down so as to compete with conventional meat producers? Will government regulations or pushback from the meat or agriculture lobby thwart progress? Is the technology close to ready? And even if the first clean meats start appearing on shelves, as optimists within the industry predict, will people even eat it? But there's another question being asked by food reform advocates—including even some who support cellular agriculture: With improvements being made in the plant-based protein sector, yielding truly meat-like products derived without any animal products at all, is clean meat superfluous?

I remember the first time I ate a veggie burger. The year was 1993. I'd recently become a vegetarian because of an ethical concern about animal welfare, and I was looking forward to trying my first burger as a newly minted herbivore. I wondered if it would be just like the meal that had brought me so much pleasure in the past. As I bit into the patty and started chewing, I recall a sense that while I enjoyed the product—a feeling probably aided by my knowledge that then president Clinton was reportedly eating the same Boca Burger brand in the White House—it didn't taste like an actual hamburger. It wasn't that it was bad—I really did like it; it just was very clearly not a hamburger.

Fast-forward more than two decades later, and plant-based burgers from companies like Beyond Meat and Impossible Foods, and chicken from companies like Gardein—none of which existed in 1993—often fool the most inveterate carnivores. Some of these brands, like Gardein and Beyond Meat at least, have strong distribution in mainstream supermarkets, yet they're often—though not always—relegated to the ghetto of the frozen natural-foods sections where most protein shoppers don't instinctually gravitate. But when people do try them, they're generally quite pleased.

In blind taste tests I've administered to friends and family, it's difficult for most to discern that these meats didn't come from an animal. Innovation in food science, much of it funded by investors like Bill Gates and Li Ka-shing, has brought plant proteins to heights the makers of the first generation of veggie burgers could only imagine. Some of these new burgers are even created to "bleed" when you cut into them, a phenomenon that makes for a great party trick, in my experience.

If you think about just how far plant-based meats have come in the past decade, it's not hard to envision what they'll be like in another decade. "Consumer awareness combined with innovation is driving the increased popularity of plant-based meats," says Michele Simon, executive director of the Plant Based Foods Association, a trade group representing vegetarian food companies on Capitol Hill. "It's exciting to imagine what the next ten years will bring as more companies raise the bar with great-tasting options."

In other words, considering just how meat-like and milk-like some of the plant proteins have become, do we really need cultured animal products? It's completely possible that by the time clean meats hit the market at affordable prices, the plant-based products may be already sufficient to offer carnivores the taste and texture they're craving.

"If veggie burgers can satiate the 95 percent of people who eat meat, great," Mark Post declares. "I'll be happy. But if that's not going to be the case, you need a backup. And this is a good backup, since it's still the real thing. It allows meat-eaters to continue their habit without the negative consequences on the planet."

It's hard to argue with Post. If meat-eaters (i.e., nearly everyone, including Post) were willing to switch to plant-based foods, there'd be no need for clean meat. Of course, that's a really big "if," and one many aren't willing to bet on.

But what if plant-based proteins get so good that people are willing to switch? After all, which is more realistic: that the nascent backup plan of growing meat in vitro will come to fruition or that plant-based alternatives will improve to the point that they appeal to even the pickiest carnivore's palate? Even some in the biotech world are less optimistic about the cultured option. This is why synthetic biologist Christina Agapakis, whose skepticism of clean meat's commercial feasibility was made clear earlier, thinks "meat substitutes . . . are much more interesting and feasible."

One point of proof that plant-based alternatives may be more feasible than clean meats is that the former are already sold in virtually every mainstream American supermarket today, as compared to clean meat, which isn't yet sold anywhere in any substantial way. And many, but not all, of the concerns skeptics raise about clean meat don't apply to most of the plant-based proteins. Yes, plant-based meats involve food technology, but nothing on the order of tissue engineering, synthetic serum, and other biotech innovations clean meat companies are borrowing from the medical world.

That said, the plant-based Impossible Burger does use a genetically engineered yeast process to make the heme that forms its "blood" in the burger. (Heme is an iron-containing molecule in blood that carries oxygen, and is claimed by Impossible Foods to be a key part of what makes meat taste so meaty.) The genetically modified yeast doesn't make it into the final product, but the fact that proteins created through genetic engineering are used at all in the process has led some biotech critics to sound alarm bells about this plant-based burger. Going even further, some natural-foods proponents oppose foods like vegetarian chicken altogether, arguing they're too processed to be considered "natural."

But still, there's just not nearly as much of a reflexive "yuck" factor with eating plants that've been made to taste like meat compared to actual meat that was grown without an animal at all, and certainly even less so for other nonmeat products like soy milk. As far as plant-based meats have come, dairy-free milks are doing even better. Already comprising more than 10 percent of the fluid milk market (compared to vegetarian meats, which are less than 1 percent of meat sales), brands like Silk compete in the dairy aisle directly adjacent to cow's milk, often at cost-competitive prices. The popularity of alternatives like soy, almond, rice, and coconut milk is so high that even Ben & Jerry's, Häagen Dazs, and Breyers now offer numerous vegan flavors of their ice cream.

The ascendance of plant-based foods has big-time investors drooling, sometimes literally. When Bill Gates first tried a "chicken" taco produced with Beyond Meat's plant-based chicken, he famously declared it "a taste of the future of food." The company has even attracted former McDonald's CEO Don Thompson to serve on its board of directors, along with social impact investors such as the Humane Society of the United States and Leonardo DiCaprio.

As promising—and fast-growing—as these companies are, not everyone's convinced that they've replicated animal meat yet. While vegetarians and meat-eaters alike enjoy Beyond Meat's products, *New York Times* columnist Nicholas Kristof wrote in 2015, "If I were a cow, I might be a bit embarrassed by Beyond Meat's meatballs and Beast Burger." The company has since come out with its next generation of burger—the Beyond Burger—which is by far its meatiest to date and is being marketed in the refrigerated meat aisle directly next to ground beef at mainstream grocers.

Beyond Meat competitor Impossible Foods, founded by Stanford geneticist Pat Brown in 2012, intends for its meat to represent the "future of food" as well. Sustained with more than $200 million in venture capital investment so far, and motivated by his conviction that "animal farming is the single biggest environmental threat on the planet today," Brown almost sounds like he's in the cultured-meat business. "We're not creating a veggie burger. We're creating meat without using animals." He himself is vegan, but his fellow vegans aren't the buyers he has in mind for his burgers. Brown's goal is to convert meat-eaters to his plant-based meats, thereby converting pastures and cropland back into forest. In his words, he literally wants to "change the way the Earth looks from space."

Just as with the lifecycle analyses the clean animal product companies are funding, Impossible Foods funded its own in 2015, comparing its plant-based burgers to those produced from actual cows. The results were as dramatic as the cultured meat life cycle analyses: Brown told *Vox*'s Ezra Klein that his burgers use 99 percent less land, 85 percent less water, and emit 89 percent fewer greenhouse gases than beef from cattle. All of this is enough to make Brown one of cultured meat's skeptics. He simply believes plant-based meats will just obviate the need for cultured meats, which is why Brown protested in 2017 that growing real meat from cells is "one of the stupidest ideas ever expressed."

As surprising as it is to see, Brown—a proponent of using biotech to find solutions to key environmental problems—agreeing with some of biotech's biggest critics about clean meat, it's of course for different reasons. What's perhaps even more surprising is who agrees with him. Even Jason Matheny, founder of New Harvest, isn't sure that cultured-animal products are more promising than plant-based companies like Brown's.

"Per dollar invested," Matheny asks, "is it better to invest in cultured or plant-based? I really don't know. At this point, if I had to bet, I might put my money on plant-based to make the bigger difference, since that industry is already so well-established." But, he argues, there's so much money and momentum in the plant-based space, and the problem of animal agribusiness is so great, it doesn't make sense to put all your eggs in one basket, so to speak. Matheny offers a third strategy to consider: investing in meat extenders—products that can enable big meat users simply to use less meat. "That might be the most efficient way to cut demand for ground meat, at least," he notes, referencing plant-based ingredients that companies can add to ground meat that enable them to use less actual meat in their recipes.

Isha Datar agrees with her predecessor at New Harvest, but adds an additional thought. "There may not be a need to see a binary distinction between cultured and plant-based products," she posits. In the near future, Datar envisions a hybridization of cultured and plant-based foods similar to what Impossible Foods is doing with its yeast-produced heme. "You can easily envision initial milks and eggs using maybe half cultured and half plant-based to help keep costs down for first offerings," she suggests. It doesn't have to be either/or; it can be both."

It's surely possible that there'll be promise in blurring the lines between cultured animal products and plant-based proteins. But the reality still remains: Even the most cheerful forecasts for clean meat indicate that ground meat is still years away from being cost-competitive with conventional animal products. Plant-based companies like Gardein, however, are already selling in mainstream grocery stores whole chicken breasts that are very hard to distinguish from the "real thing," and presumably other popular whole cuts will be replicated in plant form long before cultured chicken breasts hit the market.

It's difficult to find someone more optimistic about clean meat than Bruce Friedrich of the Good Food Institute, but when it comes to putting his resources where his enthusiasm is, less than half of GFI's considerable staff time and other resources is focused on cultured companies, with the rest in plant-based. For him, that's less of a reflection of where he thinks the most promise is and more influenced by how much more work is being done in plant-based than in cellular agriculture so far.

"We just don't know if plant-based meats will ever be viewed as the 'real thing,' no matter how realistic they become," Friedrich says. "I hope they do, and I'd love for them to dominate the market so that the clean animal products aren't necessary. But the human desire for animal meat is a strong one, and I'm just not sure that anything other than actual animal meat will satisfy the most diehard meat-eaters."

Josh Tetrick, the CEO of Hampton Creek, which he cofounded with Josh Balk in 2011 in an effort to produce plant-based alternatives to foods that traditionally require eggs, agrees, though his view is based on personal experience.

In 2014, manufacturing giant Unilever—purveyor of Hellmann's mayonnaise—sued Hampton Creek for calling its egg-free spread "Just Mayo." It turns out the FDA has a World War II–era standard of identity for calling a product "mayonnaise," and part of the definition includes eggs. Unilever quickly dropped the suit after being mocked in the press, though, in an effort to appease the FDA, Hampton Creek added "egg-free" to the Just Mayo label.

"The fight with Unilever and the FDA over what we can call Hampton Creek's mayo really taught me a lot," Tetrick says. The company, which has, up until recently, been 100 percent focused on creating plant-based alternatives to conventional egg- and milk-based foods like cookie dough, salad dressing, and, of course, mayonnaise, is now interested in entering the meat market. And while the seemingly predictable path forward would be to continue to do what they do best and develop plant-based products similar to those now being offered by Beyond Meat, Impossible Foods, and Gardein (though perhaps at more competitive prices), Tetrick sees inherent limitations in their ability to make a difference this way.

Right now, the meat-free chicken products on the market all have names making clear they're not actually from birds. Whether it's Gardein, MorningStar Farms, Tofurky, or others, they all use terms like "chik'n," "Beyond Chicken," "Chik Patties," and so on. This is for the same reason Unilever challenged Hampton Creek's calling its product "mayo"—there's a federal standard of identity for foods called chicken, and it goes without saying that if it didn't come from the fowl, it needs a different name.

"I think a great, affordable, plant-based chicken could really do a world of good," Tetrick predicts. "Maybe you could replace ten, fifteen percent of chicken consumption with it. Maybe even more. That alone would be massive. But without being able to actually call it 'chicken,' I just don't think it ends the factory farming of animals."

***

As Modern Meadow labors away to bring its leather to the market, start-ups like Memphis Meats and Mosa Meat have had the clean-meat sector to themselves—until now. Hampton Creek's Tetrick decided in 2016 that rather than stick solely with plant-based proteins, he'd go all in on clean meat, especially chicken. With his company now valued at more than a billion dollars, he intends to start spending millions of dollars a year researching in the race to commercialize clean meat.

"Let's be real," says Tetrick in late 2016. "There's no way this thing gets done without a Manhattan project. The hurdles are just so high. Everyone knows we can do this, but not without massive investment of resources. And who better to take on this challenge than us at Hampton Creek? That's how Project Jake started."

Project Jake is the at-first clandestine clean-meat project named after the golden retriever who'd been at Tetrick's side for eight years and served as a mascot of sorts for Hampton Creek during the company's first half decade. Always in the office, Jake knew no enemies, only friends. Businesspeople and job applicants alike would travel to San Francisco bearing gifts for him, which he was quite happy to accept.

"Jake was a regal presence in the office who everyone wanted to pay their respects to when they'd visit," says Jenna Cameron, partnerships manager at Hampton Creek, with a smile. "Instead of coming to kiss the ring, they came to kiss the collar. Jake wasn't a pet; he was a very special part of the Hampton Creek family."

As the company continued its explosive growth from its birth in 2011 to its billion-dollar unicorn status just six years later, Jake found his way into countless news articles and profiles. One disgruntled former employee even complained to a journalist that in the early days of Hampton Creek, Jake would regularly stroll through the experimental kitchen, where he'd steal new prototypes of cookies as they awaited taste testing.

Tetrick conceded the concern in a written response to the journalist's "exposé." "That's true," he wrote in August 2015. "Jake still has a taste for sugar cookies—although he hasn't been allowed in the lab for 2.5 years."

Lab visits were just a part of his story. Jake witnessed the rise of Hampton Creek and all the battles along the way. He was there when the company's corporate headquarters served as its R-and-D facility and lab space—all inside a small Los Angeles studio apartment. He was there when the first bottle of Just Mayo was shipped to Whole

Foods Market in 2012. And he saw Compass Group—the largest food service provider in the world—replace Hellmann's egg-based mayonnaise with Just Mayo in thousands of its cafeterias in 2014.

The company continued to expand in the wake of its labeling victory with Hellmann's to the point where Jake saw Hampton Creek branch out far beyond mayo, putting egg-free cookies, cupcakes, muffins, cakes, brownies, and pancake mixes, along with dairy-free salad dressings, on the shelves of Target and Walmart. Even its egg patties made entirely from plants are now being served at universities.

"Words can't describe the hole in our hearts when he passed away," Tetrick reflected in 2016, just a month after Jake died. "Cancer took his body away from us, but his spirit lives on, inspiring us to do more good for animals in his memory. Project Jake is for him."

Hampton Creek itself was named after Josh Balk's dog, Hampton, who had also passed away. With the birth of Project Jake, Tetrick had his eyes on a new vision for the company, no less audacious than that with which the company started.

"DESTINATION: World's Largest Meat Company By 2030." So reads the sign just inside the entrance of the Project Jake lab. A blown-up canvas of Jake wearing clear plastic lab goggles greets visitors who are allowed into the covert new lab space where Hampton Creek is beginning to amass a frozen library of farm animal cell banks. If Project Jake is successful, a switch from animal meat to plant-based meat may be unnecessary for many consumers, as you'll be able to have all the chicken you want, free of the factory farm and slaughter plant.

Even though it's only recently entered the meat-production business, the obvious draw for Hampton Creek to enter the cellular-agriculture space was the tight connection with the company mission to leverage technological advances that offer the promise of radically transforming our food landscape into a more sustainable, healthier one. Replacing eggs and milk with plants is a big part of that story, Tetrick believes, but "we haven't really fixed the broken system until we solve the meat and fish problem."

Hampton Creek brands itself not as a food company, but as a food *technology* company, and argues that its competitive edge lies in part in an R-and-D platform that enables it to screen through thousands of plant species to identify functional proteins for use as ingredients in its products. For example, a Canadian yellow pea varietal acts as the emulsifying agent in Just Mayo, obviating the need for eggs. To enable these discoveries, Hampton Creek has built a proprietary plant database that employs machine-learning algorithms and the person-power of more than fifty scientists to characterize and catalog the diversity of the plant kingdom.

So when it came to tackling the meat problem, rather than starting from scratch, Project Jake already had a robust technical infrastructure in place to benefit from. In late 2016, Hampton Creek had just completed two of its newest labs—a full analytical chemistry suite and an automation lab running large robots to significantly increase the throughput of its screening and characterization capabilities. The company had also hired dozens of new R&D team members amassing, collectively, decades of experience related to protein discovery, bioprocess development, production scale-up, and engineering, all as applicable to clean meat development as they are to plant protein discovery.

With these resources at its disposal, the launch of Project Jake would come as a shock to many who knew the company only as a plant-based mayo-maker, but was hardly a surprise to those familiar with the company's history. Hampton Creek has always considered the welfare of chickens a top priority. By creating egg-free alternatives of products, especially those used by bakers and in restaurants, Balk and Tetrick sought to reduce the impact of the egg industry.

For decades, animal advocates have been pressing the egg industry to reform its practices, primarily the confinement of chickens in battery cages—enclosures so small the birds can't even spread their wings. For more than a year, they remain immobilized, just pumping out egg after egg. "The hindrance was that going cage-free costs more money," Balk concedes. "But I knew that for a lot of food manufacturers who use eggs, they really didn't need them in their cookies, cakes, and other baked goods. We could produce those same products without eggs at a more affordable cost since plant-based proteins, like those from peas, are cheaper than eggs."

Jake's passing profoundly impacted Tetrick. To see his daily companion go from seemingly healthy to deceased in less than a month forced him to ponder his own mortality. It was only six years earlier that Tetrick himself had a near-death experience caused by a heart defect, which now prevents the CEO from engaging in strenuous exercise.

Tetrick wondered, "What would I do with my life if I knew I only had five years to live?" Replacing the egg industry would be huge, he thought; that'd be several hundred million animals spared from misery each year in our nation alone, plus enormous environmental benefits. But the real numbers of animals are in chicken meat. "Henry Ford didn't replace just some of the horses in our streets. He made them all obsolete."

By launching a clean-meat initiative, Hampton Creek is hoping to eliminate chicken suffering altogether. So much of the clean-animal-product attention until now has been devoted to replacing cattle with something eco-friendlier and more humane. For the most part, whether the research is for clean beef (Mosa Meat and Memphis Meats), milk (Muufri/Perfect Day, whom we'll meet in the next chapter), or leather (Modern Meadow), it's the cow who stands to benefit in the same way horses did from the car's invention. Yes, Memphis Meats produced clean chicken and duck in 2017, but at least until then it hadn't been the primary public focus of the company's efforts.

For some clean-meat enthusiasts, like Mark Post, the primary motivation behind the bovine focus is environmental. Conventional beef may be the most ecologically destructive commonly eaten meat in the world, so shifting humanity's diet away from bovines is extremely important to protecting our environment. But from a strictly farm-animal welfare perspective, the vast majority of the suffering caused by terrestrial factory farming is endured not by cattle but by chickens.

(If you include aquaculture as factory farming, fish rival or even exceed chickens in this respect. Fortunately for fish, clean meat start-up Finless Foods is working to multiply their cells, echoing Jesus's biblical multiplication of the fishes, ultimately aiming to bring clean fish to market. CEO Mike Selden jokes to me while I'm sampling his clean fish in September 2017 that dolphin-safe tuna isn't enough; he wants tuna-safe tuna.)

Even if clean beef replaced all conventional beef in America, the change would affect less than 1 percent of all farm animals in the country. Simply put, nearly all terrestrial farm animals in America are birds. There are thirty-five million cattle slaughtered for food annually in the United States, compared to nearly nine *billion* chickens. That means that for every one cow who enters a slaughter plant in the United States, 257 chickens enter at the same time. Put another way, when you include turkeys, that's nearly three hundred birds killed every second of every day inside slaughter plants around the clock—just in America alone.

This isn't, however, solely a numbers problem. Beef cattle spend the first portion of their lives at pasture, and even after they enter feedlots, they're still outdoors and able to walk around. But chickens, turkeys, and ducks raised for meat are typically confined inside windowless warehouses by the tens of thousands for their entire lives, and many of them suffer from chronic pain caused by genetic manipulation that causes them to gain weight rapidly and unnaturally so they can be slaughtered that much sooner.

There's little doubt that replacing poultry products with clean alternatives would do much more to reduce animal suffering than replacing cattle, though both are vitally important. "If you want to kill fewer animals and cause less animal suffering, but still want animal-based meat, switch from chicken to beef," says Lisa Feria, the CEO of Stray Dog Capital, an investor in clean-meat start-ups. "There's so much less suffering in each serving of beef than chicken. Not only are there hundreds of times more 'servings' of meat in one cow than in one chicken, but the cow was probably treated much better. Of course, it'd be better simply to leave both chickens and cows alone."

Still unsure just how best to take on the problem of replacing chicken, Tetrick began thinking and reading more about it. Plant-based milks have proliferated in the marketplace, while plant-based meats lag woefully behind. What if there was a better way?

Tetrick first heard of cultured meat in 2007, while he was in law school at the University of Michigan. "I'd been introduced to Jason Matheny of New Harvest by Josh Balk, and he sent me the NASA goldfish paper on the topic. I remember being in class and reading it. It was a cool idea that intrigued me, but it didn't cross my mind for a second that I'd one day have anything to do with it." He went on to work as a lawyer, only to be dismissed from his job after publishing an op-ed in the *Richmond Times-Dispatch* in 2009 that contained a sentence condemning the cruelty of factory farming. "It turned out that our firm represented a large meat company that didn't like that line," Tetrick says, laughing about the incident years later. "As a junior attorney, upsetting a big client wasn't helpful for my career in law, but without that op-ed costing me my job, I might never have started Hampton Creek."

Fast-forward to 2016, to Jake's death, and Tetrick and Balk were sitting in the CEO's San Francisco apartment kitchen having Chipotle's tofu sofritas bowls for dinner. Discussing the fragility of life and how uncertain any of us are about how much time we have left, Balk threw the idea out there. "You've wondered what you'd do if you had only five years left," he told Tetrick. "Well, if we think meat, especially poultry meat, is where it's at in terms of the biggest impact we could make and no one else is creating it, why not us?"

As they discussed the idea more deeply into the night, Tetrick wondered why no one else was doing it. Was it because the hurdles were too high, or just because no one was prioritizing it? Balk explained that other start-up CEOs in the space obviously want their meat to be cheaper than conventional meat, and since beef is more expensive than chicken, it's less difficult to get the price point down to beef's level. But the fact that it was harder, for *all* meat, including chicken and fish, and that no one else was focusing on it made the idea even more appealing to Tetrick. "Let me put it this way: We have resources that others don't. If it's not us, then who else could do it? Hampton Creek is about going big. And there's nothing I can think of that's bigger than this."

Today, at such an early stage in its growth, Hampton Creek is spending a colossal one-third of its budget on research and development. Traditionally, the global food industry is the stingiest among major sector spending on R-and-D, often setting aside less than 1 percent of its budget for the matter. This is in contrast to the highest R-and-D investors—like computing (25 percent), health care (21 percent), and automobiles (16 percent). The food industry has been far more focused on marketing existing products than developing new ones. But according to Tetrick, Hampton Creek's mission was never to be a "nice little mayo company." It was to fix the biggest problems in the food system today. That's one reason the company has so quickly branched out from one flavor of mayo to dozens of other products now out on the market and on conventional grocery store shelves.

Since the company has a research-oriented mentality, Tetrick wondered, what could they do if they allocated a huge portion of their research dollars into clean meat? The first thing he needed to do was assemble an initial team to figure out if this was even a feasible idea.

"If you can think of anything else I could be doing to prevent over a million life years of suffering per hour, let me know." That was Eitan Fischer's response when Balk asked him if he would like to found a new division at Hampton Creek dedicated to making clean meat a reality. Such utilitarian math is typical for effective altruists like Fischer: if clean meat will be the catalyst for the end of factory farming, as Fischer believes, every hour that moment is accelerated spares an unfathomable amount of suffering.

An Ivy League JD-PhD graduate who had previously started an animal welfare nonprofit, Fischer might not have been the most obvious choice to lead Project Jake. "He's brilliant," Balk says. "I knew he'd be up to speed and moving fast in no time." Fischer was working toward forming his own clean-meat start-up before he got the call from Balk. "It was a no-brainer," he says, "starting with a few million in seed stage funding versus joining a billion-dollar company with the infrastructure and track record to make this actually happen. I really didn't have a difficult choice to make."

But even with the resources at his disposal, Fischer is cautious not to understate the challenge. "Basically, there are a multitude of things we need to get done," he says, walking through the Hampton Creek office toward the "AUTHORIZED PERSONNEL ONLY" sign on the door to the Project Jake lab. As we enter, he pulls up an Excel spreadsheet marked "confidential" showing a host of data, including cost models for various projections. After speaking to more than two dozen scientists in the field during his first two months, Fischer is confident that with the right approach, Hampton Creek can get it done.

"When I get up in the morning, I'm thinking about avian satellite cell culture. When I go to sleep, I'm thinking about avian satellite cell culture. It's probable that I know more about these particular cells than anyone else on earth, and let me assure you: they're easier to work with than cattle or other mammalian cells." So says Paul Mozdziak, poultry science professor at North Carolina State University. His department is officially named the Prestage Department of Poultry Science, after the Prestage family, which owns Prestage Farms, the turkey and pork giant headquartered in North Carolina and with production facilities across the United States.

Even though the American start-ups in the clean-meat space aren't primarily focused on chicken, Mozdziak has devoted his entire career to proliferating chicken and turkey muscle cells. As a result, when Hampton Creek began its investigation into the cultured chicken game, Mozdziak was among the first people Fischer called.

The middle-aged professor had just received a six-figure grant from New Harvest "for the purpose of creating chicken and turkey meat without animals," and had brought his graduate student Marie Gibbons on board with the funds. Their goal is simple: to help establish an animal cell line that can be used as a common research tool by academic researchers at other universities. In essence, having avian "starter cells" available to any researcher who wants them would mean, according to Datar, that "dependence on slaughtered animals as an initial source for cells will be reduced."

Mozdziak recalls culturing cells back in 1992 with some friends, exploring various meat science applications. As the cells continued growing in their modest lab, he joked to his colleagues, "Hey, if this works, you know we could grow meat in vitro?" They all laughed the idea off, thinking it so bizarre that no one would ever want to do it. And that was it for the next decade. Despite being surrounded by avian cell culture day in and day out, the scientist never contemplated a potential food application for his work.

But the idea arose again in 2004. "I was teaching a cell culture class and some of the students let their culture grow too long, producing actual chicken muscle in vitro. They didn't really even think about the fact that they'd just produced 'meat' per se; it was just muscle to them. But I thought about it, and wondered just how much like chicken meat it'd look like if we just let it keep growing." The professor didn't consider tasting it himself, and since there was no funding for such a project, he threw the little morsels of meat away. "At the time I just couldn't imagine that anyone would have an actual interest in doing something like that. How wrong I was!"

Now more than two decades after his initial thoughts on the topic, Mozdziak knows it's no longer a laughing matter. In addition to his New Harvest grant and Hampton Creek knocking on his door to learn what he knows, he's also had conversations with Tyson Foods and others he describes as having "a real interest in this kind of disruptive technology."

Part of his message to these companies is that the focus on growing cattle and pig muscle cells to produce beef and pork is noble, but just from a technological standpoint, Mozdziak argues that chicken and turkey cells are much easier to work with. "First things first, they grow a lot better in culture than mammalian cells do. They have better plasticity—you can get them to do what you want much more easily." Interestingly, he's not sure why, but Mozdziak points out that with mammals, it's easier to work with cells biopsied from younger animals, whereas with birds, a more mature animal has better satellite cells with which to start.

There are key innovations in cell culture the professor thinks will inevitably occur in the next couple of years, like not using any antibiotics in the culture, and going serum-free. But the bigger issue he sees is simply setting up immortal cell lines so that other researchers can more easily start solving the big problems.

When asked how his colleagues at the Prestage Department

of Poultry Science react to the fact that his work could put poultry farmers out of business if he's successful, Mozdziak shrugs his shoulders with a smile. "They really think of me more as a biologist than a poultry scientist, and I'm not sure how many of them are that familiar with what I'm doing yet. But those who do know about it think it's really cool."

Already, Mozdziak's New Harvest grant has yielded big gains. He used the funds to support Gibbons in trying her hand at something no one had ever done before: culturing turkey meat. For Gibbons, a physiology graduate student and lifelong animal advocate, this project was a dream come true. Sporting an oxytocin molecule necklace and a multilingual "love" tattoo, she explains how she's always been crazy about animals. "I was raised on a small family farm in North Carolina, a little over an hour away from the largest slaughterhouse in the world. Unlike most farms however, all of my family's animals were raised for pets, not profit. I loved my chickens and turkeys just as much as I loved my dogs and horses, and I still do!"

As a teenager, she learned about the animal welfare and environmental concerns brought about by animal agribusiness, leading her to write off all animal products for good. Given her love for animals, as well as an interest in science, Gibbons decided to attend North Carolina State University and pursue a degree in veterinary medicine. "To be honest, I was never quite sure if veterinary work was right for me. But I loved animals, and I loved science, so it seemed like the best option."

In preparation for her veterinary studies, Gibbons began working with a large animal veterinarian, traveling around to local farms and treating a variety of different species. "Working on these small family farms, the majority of which were organic and even certified by animal welfare groups, really opened my eyes to the differences in treatment between farm and companion animals. Just because these animals were free to graze in a pasture doesn't mean they weren't castrated, dehorned, and branded, usually without pain medicine or veterinary supervision."

Gibbons makes a key point. Advocates for a return to small-scale, organic animal agriculture often paint a picture of the "good old days" before factory farming, and set up a dichotomy between big versus small ag, with of course big being bad and small all but romanticized as the embodiment of human harmony with nature. The reality is quite different, with many abuses, like those enumerated by Gibbons, being prevalent even before factory farming ever became the norm.

And the problems for wildlife near pastured farm animals can be substantial, too. Ranchers whose cattle spend their time grazing are often those spearheading the lobbying efforts for shooting wolves in the United States and for rounding up wild horses. Many of them simply don't want predators near their cattle or competitors for grass on the federal land their cattle occupy. It's not that pasturing animals isn't better than keeping them locked on factory farms—it's immensely better for the animals—but it would be a mistake to conclude that local, organic animal production is free from animal welfare concerns.

Gibbons became convinced that she didn't want to be a veterinarian after being called on as a student to perform an eye removal surgery on a fully conscious cow at a pasture-based operation. "I decided that as a veterinarian, I could help thousands of animals by promoting humane treatment and educating farmers about proper veterinary care. But as a cultured-meat scientist, I could prevent the suffering of billions of animals by sparing them an awful existence."

Working with Mozdziak, in late 2016, Gibbons grew the first-ever cultured turkey nugget, and for only $19,000 (a giveaway compared to Post's $330,000 burger). Perhaps even more impressive is that Gibbons can send any academic scientist a vial of her starter cells—named the MG1 line after her initials—and they'd be able to grow their own turkey nuggets in just two weeks. For perspective, turkeys on factory farms typically take fourteen to nineteen weeks to reach slaughter weight.

If the cells were able to proliferate in optimum conditions, the potential amount of meat they could produce would be astronomical. Gibbons performed a biopsy by removing a piece of turkey muscle the size of a sesame seed, which contained roughly twelve million satellite cells. After some simple calculations and measurements, she and Mozdziak found that with sufficient production capacity, there'd be no reason a single biopsy of that size couldn't theoretically produce enough turkey muscle to supply the current global annual meat demand (if we were content to eat no meat other than turkey) for more than two thousand years.

To put it another way, when *MIT Technology Review* featured Mozdziak and Gibbons's work in late 2016, it reported, "In theory, the growth potential is enormous. Assuming unlimited nutrients and room to grow, a single satellite cell from one single turkey can undergo seventy-five generations of division during three months. That means one cell could turn into enough muscle to manufacture over twenty trillion turkey nuggets."

Gibbons smiles thinking of the potential for her work. "Of course, there's still an awful lot to be done in order to optimize this system. It's not like I have a few billion tons of cultured turkey meat in a freezer somewhere." She lists the barriers still facing her work, including finding a sustainable animal-free medium (which she believes is inevitable in the near-term), adapting the cells to bioreactor culture, and figuring out how to scale up production systems. And then there's the issue of what to call the final product. "I for one am a fan of calling it my bootleg turkey leg!" Gibbons adds jokingly.

To Gibbons, efficient cultured-nugget production could lead to endless possibilities. She's excited about the positive impact clean meat will have on farm animals, but even more broadly on the entire animal kingdom as a whole. "Once we stop relying on animals for food and profit, I believe that society will start treating all animals with more respect."

Gibbons's work has generated a lot of headlines and certainly earned the respect of the Hampton Creek R-and-D team that reached out to them to see if they might be able to collaborate. But the MG1 line is owned by NCSU and cannot be used by a for-profit company without a license (typically for a hefty up-front fee or royalties). Until that happens, at least, her work will remain in the realm of academia.

For Mozdziak, this is just the next step in improving efficiency, which poultry scientists have been working on for generations now. But his argument goes far beyond efficiency gains, and appeals to the sci-fi fantasies of so many of his colleagues in the biotech sector. "One thing they always agree with me on is that this is the only way we're going to feed long-term astronauts with meat. You think you're going to travel the cosmos carrying Noah's Ark with you? It ain't happening." Protein needs of long-distance astronauts may not be at the top of the list of problems humanity faces right now, but it's surely on the mind of some futurists. "If human space colonizers want meat, it's almost certainly coming from some type of reactor they'll be carrying on board, and this research is the beginning of making that happen."

Yet the reality remains: We'll know long before humans are colonizing the solar system if Project Jake is a success for Hampton Creek. And if it is, the poultry farmers supplying Prestage, Pilgrim's Pride, and other big agribusinesses will indeed find themselves out of luck. As the New Harvest press release announcing its funding of this NCSU work points out, "the outcome of the project will greatly reduce the numbers of chickens and turkeys on which humans rely for meat."

Over at Hampton Creek, as Project Jake got started, the research challenges quickly became clear. "It has to be serum-free, for starters. That's step number one," Fischer says, acknowledging that the use of any animal-based blood serum to feed cells is really a nonstarter for both ethical and financial reasons. He goes on to describe the need for cheaper nutrients to feed those cells. "More importantly, we need growth factors or mimetics that are dramatically less expensive than now. It's one thing to use them for medical tissue engineering. But for a commercial food application, we just have to bring the cost way, way down."

To solve that problem, Hampton Creek has at its disposal a patented plant-based technology platform, nicknamed Blackbird, which it intends to leverage toward clean meat. (A stuffed actual blackbird hangs suspended over the project area in the Hampton Creek headquarters.) The tools they'd built over the course of the previous three years could enable solutions to some of the most difficult technical challenges in clean-meat development. In addition to plant-based scaffolds for cells to grow on and phytomimetics—plant-derived substitutes for expensive growth factors that are essential for cell growth—an important advantage of Hampton Creek, it argues, lies in its position to develop an animal-free nutrient feed ("media") for the cells.

"There's no question," Fischer explains, "without cost-effective, animal-free culture media, clean meat will never happen. Period." Despite all the attention clean meat has received in recent years, without this, it may well remain a pipe dream in the minds of those who seek a better future for animals and the planet, unless an abundant and cheap source of nutrients to feed the cells is discovered. Looking at commercially available animal-free media today, the critics of clean meat would be undoubtedly correct in concluding that current costs of production render it a prohibitively expensive luxury product at best.

But Hampton Creek plans to change that. Project Jake, like the rest of Hampton Creek, is not meant to create a high-end product for the lucky elite who can afford it. Tetrick's mission, just like those of the other clean meat purveyors, has always been to make sustainability competitive on cost. And as it happened, the CEO asserts that if there was one company well positioned technologically to discover cheap, animal-free protein sources that would make scalable clean meat a reality, it was Hampton Creek. With a proprietary database containing the functional and molecular properties of hundreds of plant proteins, and the ongoing high-throughput screening of hundreds more, the company asserts it has at its disposal the arsenal of physical and digital data required to identify which proteins can make cells grow.

And as with its ingredient discoveries, Tetrick's plan is to make scientific advances that will not only allow Hampton Creek to create new food, but to disrupt the entire food system by enabling others to do the same. Just as the company licenses its plant proteins to the giants of the food industry to enable them to make their own products egg-free, Tetrick intends to leverage its discovery of plant-based media sources to enable other clean meat companies, as well as hopefully—one day soon—the Tysons and Perdues of the world. In other words, Hampton Creek wants its competitors to be producing affordable clean meat, too, with its technology of course.

Although surprising to some, Tetrick explains about the large meat companies they've been in discussions with that "they get it more than most believe—they've just been saddled with limited tools and a limited mind-set." Reflecting on how outdated current meat production seems by comparison, Tetrick says, "Animal ag has turned animals into meat-producing factories. The problem is that they're still feeling machines, and really inefficient machines at that. What we're doing is the same, but so much more efficient, sustainable, and humane. And the end product will be exactly the same: it's meat."

Well, maybe. Tetrick has his eyes set not simply on making chicken and other meat products, but on making them better. "The main difference between our meat and conventional meat is that conventional meat will be a hell of a lot more likely to make you sick."

In 2014, *Consumer Reports* announced its survey of three hundred chicken breast samples purchased from American grocery stores. The result: virtually all of it—97 percent—harbored dangerous bacteria like *Salmonella*, *Campylobacter*, and *E. coli*. "More than half of the samples contained fecal contaminants," *Consumer Reports* warned. "And about half of them harbored at least one bacterium that was resistant to three or more commonly prescribed antibiotics." Wonder why we hear incessant warnings about thoroughly cooking poultry products and keeping them away from other groceries? It's because they have feces on them.

The sobering survey continued to describe how, in part because of widespread abuse of nontherapeutic antibiotics in the American poultry industry to make chickens grow more quickly and keep them healthy in such unsanitary living conditions, some of our most important antibiotics are being rendered useless for humans. "Antibiotic-resistant infections are linked to at least two million illnesses and twenty-three thousand deaths in the United States each year," *Consumer Reports* noted. This is one reason the Consumers Union—which publishes *Consumer Reports*—lobbies so strongly, yet so far unsuccessfully, to ban nontherapeutic antibiotic use in animal agriculture.

But as worrisome as antibiotic resistance is, the more immediate threat from the poultry industry's practices is food-borne illness. Sadly, forty-eight million Americans are sickened annually from food contaminated with *Salmonella* and other pathogens. And the biggest source of the problem is chicken and turkey meat. "More [food-borne illness] deaths were attributed to poultry than to any other commodity," reports the Centers for Disease Control and Prevention (CDC).

"Just think how amazing it would be to have chicken meat that you didn't have to treat like radioactive waste in your kitchen," Tetrick exclaims. "Of course we won't have any *Salmonella*, since that's an intestinal bug. Well, guess what? We won't need intestines to produce our meat. People one day will marvel at how we thought it was normal to essentially play Russian roulette every time we ate some meat."

In parallel with Marie Gibbons's early advances, a start-up in Israel also realized the need for and potential of pursuing cultured-meat research focusing on birds. Founded in 2016, SuperMeat's ultimate aim is, like Hampton Creek's, to produce clean chicken. But rather than going the same route as the other companies and seeking venture capital, SuperMeat's first task was creating a viral video about the birth of their company and the promise their slaughter-free meats could bring.

The production, fast-paced and comedic, was anything but a boring science video. In fact, within a month it'd racked up more than ten million views and raised a quarter million US dollars from five thousand donors, mostly outside of the Holy Land. Since the donors were promised vouchers for any future products the company may make, SuperMeat CEO Ido Savir jokes that these people are the first to ever purchase clean meat.

With a successful viral video behind Savir, his team—including Koby Barak and Shir Friedman—got to work, seeking to use their newfound funds to prove that they could create an actual product that would attract real seven-figure investments. Already they've attracted the attention of some of Israel's biggest meat companies, with one, Soglowek publicly expressing interest in investing.

At the same time as SuperMeat was just starting to seek such investment after taking the Internet by storm, over at Hampton Creek, Fischer and the second member on his team, David Bowman, were coming up with their initial research plan. "What if we could get to market quickly with a technically easier-to-make product, but also one which would be a high-end luxury product that chefs and foodies everywhere would want to get their hands on?" Bowman asked the team. He'd worked with liver cells before coming to Hampton Creek, and suggested that foie gras (fatty duck liver) particularly stood out: the delicacy product is marketed at such a high price that getting a clean version of it to be cost-competitive would be less difficult than trying to compete with commodity chicken at first. Tetrick and Balk were intrigued.

"It's a poultry product," Fischer agreed, "making it a good candidate to start with, since the cell lines and media conditions would be relevant to similar products we want to make such as other duck meats, chicken liver, and other poultry products."

The process, it turns out, really is easier than culturing muscle cells in vitro, as liver may be easier to grow without serum than muscle, which would slash the cost of production. Moreover, if you feed liver cells too much sugar, they get fattier and fattier to the point where they mimic the hepatic lipidosis that's induced in ducks and geese when they're force-fed to produce the delicacy.

The dish has been a cultural flashpoint for years in the debates over animal welfare. In order to coax the bird's liver to become fatty, producers must use a pipe to force-feed the birds daily more than they'd ever eat naturally, causing the liver to balloon up to ten times its normal size. By the end of the process, mortality rates are sky-high and many of the survivors are so bloated they can barely walk.

Hampton Creek wasn't the first to contemplate producing foie gras in vitro. In 2008, *New York Times* columnist Andrew Revkin wrote a blog entitled "A Path to Fowl-Free Foie Gras?" in which he made the case that, considering how contentious the battles over foie gras between foodies and animal welfare supporters had become (Chicago had just banned the sale of foie gras, and then lifted the ban), why not keep the liver and lose the bird? "I think foie gras could be the perfect test case, all you cultured-meat entrepreneurs," he declared. (Of course, there weren't yet any actual cultured-meat entrepreneurs at that time to whom he could have been referring.) Revkin's argument was essentially that liver cells are easy to work with and foie gras is already quite pricey, so competing on price would be easier.

Hampton Creek's goal isn't to focus on foie gras just to displace the fatty liver industry. Instead, it's to build a complete technology platform, one that would enable the developments of countless products, and especially poultry.

But the path is still far from clear. The first question is whether actual foie gras consumers—people who may typically place a high value on what they perceive as an artisanal food—would even want to eat a lab-produced version. (It's the same problem for the lab-grown diamond makers: a cultured version, even if identical to those mined from the earth, just doesn't carry the same connotation.) Fischer is optimistic, though: "Foie gras is currently produced in such an inhumane way that our product will provide an opportunity to enjoy this delicacy without the accompanying ethical concerns."

It's possible there are some foie gras consumers who'd prefer a cruelty-free version, but most people eating foie gras are aware of the controversy surrounding its production and yet still order it. It's easy to see people not caring where their chicken nuggets came from as long as they're tasty, safe, and affordable, but many foie gras consumers may have as much fidelity to the "heritage" or tradition of foie gras as they do to its taste alone. Because of that uphill battle, Hampton Creek will have to ensure its fatty liver is truly on par with the traditional version, lest it appeal solely to those who today wouldn't consider eating force-fed foie gras on ethical grounds.

Foie gras is graded by its iron content and number of blood vessels (the less of each, the better). As a result of its production method in which they can control these factors, Fischer projects that Hampton Creek's product will be "the world's highest-grade foie gras."

Whether this will be enough to revolutionize the $3 billion global foie gras market is yet to be seen. But even if the product won't be a best-seller anytime soon within the niche category, Fischer estimates that going this route will hasten the commercialization of the first clean-meat product, something that would not only generate major attention for Hampton Creek, but could also attract more research dollars as people see that commercialization is no longer theoretical.

And then there's the added value of engaging in the foie gras business in California. Due to animal welfare concerns, then-governor Arnold Schwarzenegger terminated the practice in California, signing a 2004 bill that phased out production and sale of foie gras from force-fed birds by 2012. The law underwent many litigation challenges, but in September 2017 a federal court upheld the ban, meaning it's illegal to sell foie gras in the Golden State if the birds were force-fed. "What if Hampton Creek could be the only company to legally produce and sell foie gras in California?" Fischer only half jokes. "Think that'd get some headlines?"

_____

As I sit in the Hampton Creek kitchen a few months later in January 2017, a team of scientists is already hard at work behind me making the world's first clean foie gras, while developing cell lines of various other species. One of those scientists is Aparna Subramanian. A stem cell biologist with fifteen years of experience, Subramanian commutes to San Francisco every week from LA, where her husband and children live, to spend her time growing and feeding farm animal cell lines. Two months prior, Fischer had reached out to her on LinkedIn, and she at first thought it was a joke. "It's the craziest thing I've ever worked on," she says. "I didn't think this was for real." As a vegetarian and a deep believer in the company's mission, she describes an experience from the early days of the Project Jake lab. "To establish a new line, we have to source the starting material—the cells—from actual birds." Partnering with a local pasture-based farm to identify the highest quality birds to painlessly source the cells from (there are even stem cells at the root of a detached feather, she explains), "Ian" was the lucky chicken chosen for the task. "We got Ian from the farm and brought him to his new backyard home where, rather than being slaughtered at only a few weeks old"—Subramanian tears up—"he'll live out his life as a chicken should." Meeting Ian, she recounts, "really reminded me that despite the many late nights in the lab, this is all worth it. We're doing it for them." Since she is responsible for making some of the world's first animal-free meat, the irony is that Subramanian—whose vegetarianism is religiously based—won't get to taste the fruits, or meats, of her labor. "It's real meat," she explains, "an animal product."

After a successful few months of research by Subramanian and the team, Project Jake is no longer just a project. With the expertise of Hampton Creek's R-and-D scientists, such as Jason Ryder, who previously oversaw the manufacturing of four-hundred-thousand-liter bioreactors, and Viviane Lanquar, who'd cataloged the molecular properties of hundreds of plant proteins that may be used as media ingredients, clean meat is now a central pillar of the company's strategy. Its two approaches focus on the "seed"—harnessing the power behind the diversity of the plant kingdom—and "cell"—leveraging the exponential ability of animal cells to proliferate, made possible by the discovery of plant-based media ingredients.

With some early breakthroughs, Hampton Creek is now ready to begin taste testing its initial samples made with no animal serum, and I'm the first person outside of the company to get to try them. Thomas Bowman, the lead chef on Project Jake, had, along with his brother David, been imagining this kind of moment for nearly a decade. The duo daydreamed about the possibility of making cultured foie gras since Thomas started cooking professionally and David began culturing liver cells, and now they were about to debut their fantasy product to an outsider.

"Foie," as Thomas (and foie gras connoisseurs) calls it, "is regarded by top chefs as perhaps the most prized animal product today." The early prototype—which just like typical foie gras mousse is composed primarily of fatty duck liver, together with other ingredients—that he serves me looks like and smells of duck liver as far as I can tell. "This *pâté* de foie gras," he explains, "can go for up to one hundred dollars per pound in retail." Fischer adds: "We already make it in a scalable process; it's just a matter of time until the price points are where we need them to be." (Producing muscle meat serum-free will take more time, but as noted, the liver is easier to work with.) Solving a key problem in the cultured meat challenge—how to make meat without a continuous supply of nutrients from animals who needed to be raised and slaughtered—the Hampton Creek team was able to leverage its knowledge about the functionality of animal media ingredients to find viable animal-free replacements, much like the company had previously done with egg ingredients in mayo and cookies.

I've never eaten foie gras in my life, and I'd campaigned a decade earlier at the Humane Society of the United States to ban its sale, on animal cruelty grounds, in Chicago. When California's foie gras ban took effect in 2012, I regularly did public debates with chefs who defended their product. Their vehemence reminded me of George W. Bush speechwriter Matthew Scully's quip about such foie gras defenders, pondering just how "a man rising in angry defense of a table treat has any business telling other people to get serious." Yet here I was about to seriously consume actual foie gras.

Its beige color stood out on the white ceramic plate before me, fork and knife on either side with a high-end napkin; it looked almost as if I were in a fine French restaurant. As I sat down with a crowd of Hampton Creek staff watching and awaiting my reaction, I felt my stomach churn just a bit at the thought of what I was about to do. Knowing that no duck was actually slaughtered was sufficient to persuade the rational side of my brain, but my visceral reaction was still intense. As with the steak chip I'd eaten a couple of years before, without much fanfare, I cut a piece of the foie gras with my fork, raised it to my mouth, took a breath, and slowly pressed the foie gras with my tongue against the roof of my mouth.

The flavor was impressive. The *pâté* was rich, buttery, savory, and very decadent, just as one would expect. I'm certainly not the best judge in this case, but as I closed my eyes and let the fatty liver melt on my tongue, the Hampton Creek foie gras brought me an amount of pleasure I'll confess I was a little embarrassed to admit. There's just something about fat that really makes the human brain happy. My typical experience in that realm would be compared to eating a fatty food like guacamole—which certainly does produce a lot of happiness—but this foie gras was in another league altogether.

As other members of the Hampton Creek culinary team joined in to taste the newest iteration, reactions varied from surprise to relief. Fischer joked, "I had protested foie gras as a college student, and here I am, eating it every other week." Even with my positive reaction, the team pressed back that it still wasn't where they wanted it to be. Since this may end up being the first cultured meat to ever be commercialized, anything less than total perfection would be a letdown for them. "Until it scores better than the force-fed version on our blind tests, not a single consumer will buy this product," says Thomas.

Still, Tetrick keeps his eye on his biggest prize. After tasting some of the company's clean foie gras after I'd had my sample, he declared that "foie gras is great. It's an achievement we'll be proud of and perhaps it builds a bridge. But we know where that bridge is leading. We want to render the current model of meat production totally obsolete."

On the screen, he flips through some of Hampton Creek's drawn-up models for a future four-hundred-thousand-square-foot meat-production facility: two hundred bioreactors, producing seventy-six pounds of bluefin tuna per second, alongside clean Kobe beef, and—ultimately—what he boasts will be the best chicken meat the world has ever seen. "Our goal is to make this stuff so obviously better, that there'd never be a reason to choose the conventional kind." Showing off their proposed timeline, he tells me, "By 2025, we'll build the first of these facilities, and here"—he points to 2030—"we're the world's largest meat company."

The plan right now is to make the first sale of an animal product made without requiring the use of an animal by 2018, at a price "within shouting distance" of the conventional product, says Tetrick. When I press him as to what he defines as "shouting distance," he offers 30 percent higher as a goal for the initial commercial offerings. But he insists the problem isn't solved until clean meat undercuts the prices of meat today. To do that, Hampton Creek has more work ahead of it to reduce the costs of its plant-based media. Within the next five years, Tetrick estimates, they'll get there. Already they've produced clean chicken nugget prototypes, and consumed them—as depicted in a short film they produced—in Ian's presence.

I asked Tetrick how he can be so optimistic. "Look, we already have 7.5 billion people mouths to feed on the planet." Tetrick pushes back against theoretical opposition to commercializing his ultimate product. "They need to be fed, and so do the coming billions who'll soon be here. I'm quite confident that huge developing nations—especially since this is going to solve so many food-safety problems—are going to be very happy to have our product when it's more affordable than their current meat. Even if the EU doesn't want it at first, countries like Israel, China, and Brazil will. After all, what we'll be selling is just chicken."

The CEO pauses for a second and contemplates. "That's actually a good name for it: Just Chicken. I think Jake would've liked that."

7

BREWING FOOD (AND CONTROVERSY)

Among those who recognize the massive problems animal agribusiness poses to the planet, the debate around solutions has largely focused on whether to replace the products of the factory farm with plant-based alternatives, or to invest in animal products that are grown from biopsies of actual animals. Yet there's an entirely other field that doesn't fit so neatly into either of those categories, and it's closer to commercialization than the clean meat purveyors are. In fact, in some ways it's already commercialized. In the field of cellular agriculture, these entrepreneurs are focused on the subfield of what's typically called *acellular* agriculture.

Cellular ag is best known for generating living cells (like muscle or skin cells) that can proliferate and become food or clothing. Acelluar ag entails coaxing living, microscopic organisms, like yeast, bacteria, algae, or fungi, to produce specific organic molecules such as fats and proteins that aren't actually alive themselves. In acellular ag, since you're starting with just yeast or some other microorganism as opposed to a biopsy from an actual animal, there's no animal involved in the making of these animal products. At the same time, despite no animal being involved, the proteins these companies are creating are the same exact proteins found in the animal products they're seeking to displace.

Another way to look at it is that clean-meat producers are producing foods that are made *of* cells, whereas the acellular products are foods that are made *by* cells.

Acellular food companies are embarking on a journey so different from that of companies like Memphis Meats, Hampton Creek, and Mosa Meat that they don't even like to be included in the same category. For them, the beer brewery comparison is far more applicable to their work than to the meat-makers, since just like with beer, they also simply start with just a microorganism that's designed to produce the product you want. But instead of working with brewer's yeast to produce alcohol, these companies get their yeasts to churn out milk, egg, or collagen proteins, for example.

Just as baker's yeast and sugar together converts into CO_2 to leaven bread, and brewer's yeast and sugar together convert into alcohol, these companies are using their own specialized yeasts that they feed sugar, converting it into proteins found in milk and egg whites. One key difference, however, is that in beer brewing, the yeast remains in the final product, whereas these new companies separate the yeast cells from the proteins they create, leaving just the pure milk or egg protein. Since their yeast, which is genetically engineered, is removed and not present in the final product, these foods are GMO-free. Fans call it "enhanced fermentation" (or just fermentation); opponents call it synthetic biology, or even GMO 2.0.

The culturing processes these acellular companies are using have been around for far longer than the meat-culturing techniques used by the other companies already profiled in this book. For example, until a few decades ago, nearly all the insulin used by diabetics came from pig or cattle pancreases, but in the late 1970s scientists figured out a way to genetically engineer bacteria to produce human insulin in vitro. Today, nearly all diabetics use this lab-created insulin, which is much safer for the patient, and their bodies don't even know the difference. Similarly, until 1990, all hard cheese contained calves' stomach lining, known as rennet, in order to make the dairy curdle. Today, nearly all rennet is produced in vitro, where genetically engineered bacteria serve as factories for producing chymosin, the key component of rennet. After the genetically engineered bacteria does its job, it's discarded, leaving the chymosin isolated and allowing the cheese to be labeled GMO-free.

There are limitations to such efforts (it's unlikely that such a process could produce clean meat, for example), but these products still nevertheless are far closer to being identical to animal products than the plant-based alternatives to them are. Take, for instance, Perfect Day.

Inside a South San Francisco office at the end of a cul-de-sac, just eleven miles from Hampton Creek's headquarters, a crowded communal office space is home to a dozen employees, all working for Perumal Gandhi and Ryan Pandya, two Indian-American vegans in their midtwenties aiming to leave cows out in the pasture by culturing real cow's milk from yeast. Under a microscope in the expansive lab above their modest ground-floor office space, it's hard to tell the difference between the milk Perfect Day produces and what the company calls the "cow-derived milk" at least some human cultures have been drinking for millennia. (The name Perfect Day comes from a curious study showing that bovines who listen to soothing songs like Lou Reed's "Perfect Day" produce more milk.)

While tissue-engineering meat is a massive technical challenge, producing liquid milk is far simpler, involving no tissue engineering at all. It primarily involves just a half-dozen key proteins that are remarkably easy, relatively speaking, to produce from scratch.

There's a lot to like about cow's milk, they argue, but some things would make it better. For example, not having any cholesterol would be a start, considering that cholesterol adds no taste or texture to milk anyway. Noting that the majority of humans have never evolved lactose tolerance and therefore have difficulty digesting milk, they intend to leave the lactose out, too. Another benefit: no bacteria, meaning milk with a much longer shelf life.

Perfect Day utilizes the same process as the insulin and rennet manufacturers, with its yeast (which they've named Buttercup) spitting out milk proteins instead of rennet components. "And then, once we have the milk the yeast created," Gandhi deadpans, "we slaughter the yeast cells by the trillions without even shedding a tear."

Many people may give so little thought to dairy consumption that they assume that, since cows happen to produce milk, it's natural for humans to consume it. The reality is that cows, like all mammals, only lactate when they've been impregnated, and a cow's milk is designed for her calf, not humans. (Humans are the only species that drinks the mother's milk of another species, let alone drinks milk into adulthood.) In order to keep the milk flowing, dairy farms use artificial insemination to keep cows in a constant state of impregnation and milking, and separate the mother from her calf on the same day the animal is born so they can sell the milk. The calves are then either used to replace their "spent" mothers on the dairy line once they've reached puberty (at least a year old), or are raised for veal or beef.

Intensive genetic selection programs mean that today's cows now produce many times more milk than their ancestors, a phenomenon exacerbated by the routine use of hormones and antibiotics in the dairy industry, forcing the cows to produce even more. Such high rates of milk production are associated with greater incidences of all types of animal welfare problems, including lameness and mastitis, a painful udder infection.

The introduction of cow's milk to humanity's diet is so new in our evolutionary past that, as Perfect Day points out, most humans haven't evolved the ability to properly digest lactose (they're lactose-intolerant), which is among the reasons you don't see a lot of dairy in the traditional cuisines of non-European-inspired foods. But Europeans have such a long history with consuming unfermented cow's milk that many evolved the capacity to digest lactose beyond infancy and have no problem drinking milk through all stages of life.

Ryan Pandya, who adopted a vegetarian diet as a teenager growing up in Connecticut, learned about these problems in the dairy industry after reading *Eating Animals* by Jonathan Safran Foer during his college career. "I started realizing that the same conviction to avoid violence against animals that led me to become vegetarian also necessitated that I become vegan," he recalls.

At the same time as Pandya was growing up on the East Coast, a sixth-grade boy leaving his school in Mumbai was handed a brochure about the mistreatment of chickens. Perumal Gandhi could hardly believe his eyes. He saw birds who'd been genetically manipulated to grow so big they could barely walk. Other chickens were locked in overcrowded cages. Born to vegetarian parents but raised as a meat-eater, the child instantly knew he wanted to be more like the elders in his family and swore off meat for good.

Gandhi's compassion for animals was so intense that his father found ways for him to be around animals more. Eventually they heard of the Welfare of Stray Dogs, a nonprofit organization that cares for Mumbai's many street dogs, and Gandhi began spending his free time with them.

Meeting other animal-welfare volunteers brought Gandhi into contact with people who called themselves "vegans," a term he'd never heard. But seeing the fate of India's dairy cows was enough to persuade the high school student that going vegan was the right thing to do. "Cows are sacred in India, which means Hindus don't kill them, but they still suffer enormously and are slaughtered for food anyway, just not by the Hindus." In fact, India is tied with Brazil as the largest exporter of beef in the world.

As he readied himself for college, he knew his parents would only be content with two potential career paths: Gandhi could become a doctor or an engineer. He chose the latter, but he couldn't forget that brochure he'd received so many years ago. "I began feeling a sense of helplessness. Like I could only help one animal at a time, when there were billions who needed my help. I just wanted to make a bigger impact. But I didn't know what to do."

———————————

Back in New England, in 2013, Pandya was now done with school and starting his career in biomedicine, beginning to apply the promise of biotech to medicines and treatments that would hopefully help people lead longer and healthier lives. But he was unsatisfied. As an undergraduate, Pandya had worked on early attempts to culture bovine stem cells to make animal-free meat. He knew the idea was out there and was frustrated that he wasn't advancing the field.

Pandya knew about plant-based start-ups like Hampton Creek and Beyond Meat that were making egg-free products and meat-free chicken, respectively, and how they were employing food scientists to help them make plant-based eating more convenient and familiar to the masses. But what he really missed wasn't eggs or meat—it was dairy.

"I'd be fine never eating meat again in my life," he says. "But I just love dairy so much, and I knew the dairy industry just wasn't aligned with my values."

Then one day, in what Pandya calls his Bagel Revelation, he ordered a bagel with nondairy cream cheese from a Boston deli. "As I removed the brown paper packaging, this runny, sorrowful-looking cream cheese substitute just dripped out of the bagel and onto my jeans, staining my right pant leg with this ugly gray streak. It was sickening to look at. I couldn't even eat the bagel."

Disgusted, Pandya began envisioning what was happening in that cream cheese at the molecular level. "I could almost picture this whole network of milk proteins that were just missing from the crappy soy-based spread, and I started wondering if the technology I used to make medicine at work could be used to make milk proteins." Pandya began getting really excited when he realized that it'd be much easier than tissue-engineering meat, and when a deep Internet search revealed that no one on earth seemed to have had this idea, he felt in his bones that he had to take action.

He wasn't sure how to get started, but he knew someone who'd be able to help. While at Tufts, he'd worked under Professor David Kaplan, attempting to culture meat from animal cells. For background, he was encouraged to read from the online library of New Harvest, by this point being run by Isha Datar from her home in Canada. Pandya was so interested that he kept in touch with Datar, knowing—or hoping—she'd play an important part in his career one day. As soon as Pandya realized the huge potential of his idea, he wrote a proposal and sent it to Datar, asking her if she thought there was a company in his concept.

Meanwhile, Gandhi was struggling with the question of how to best apply himself to help not just one street dog at a time but vast numbers of animals. As an engineering student in India in 2012, he'd gotten subscribed to New Harvest's e-newsletter and read about the work that Mark Post was doing in the Netherlands to bring about cultured hamburgers. "Real burgers without cows? I knew instantly that I had to get into this field."

Gandhi contacted Post, asking if the Dutch professor would take him on as a PhD student. To his great chagrin, Post simply didn't have

the resources to accept him. Undeterred, Gandhi quickly began applying to graduate schools in the United States, getting into many,
and eventually choosing Stony Brook University in upstate New York.

Despite his admiration for Post's work, Gandhi wondered why all
the attention was on culturing meat rather than other animal products. Presumably, he figured, it'd be much easier to culture something
structurally simpler than meat, like liquid milk.

In a twist of fortuitous fate, just weeks after Pandya reached out
to New Harvest, in March 2014, Gandhi did the same thing. Datar
connected the two, telling them they both had the same idea, and let
them know about a fledgling start-up accelerator in Ireland, called
SynBio Axlr8r, where biotech entrepreneurs could receive $30,000
and free lab space to explore an idea for a new company.

The New Harvest CEO proposed an idea of her own: Why don't
the three of them create a new company to produce cow's milk from
yeast and apply? They figured they had a great way to differentiate
themselves from the meat work for which Mark Post had recently become famous. The big difference between what cultured-meat purveyors were doing and what they wanted to do, of course, was that
these hopeful entrepreneurs knew they wouldn't need animal cells at
any point in the process, nor would they even need tissue engineers.
They knew they could persuade yeast to produce any type of protein
they wanted. And since milk is simply a liquid comprised of different
proteins, fats, and sugars, why couldn't they just build the milk from
the molecule up?

When the email accepting their application came back, Pandya
was so elated that the first thing he did was call his mom. "Are you
sure this isn't a scam?" the mother questioned her son from her Mil-
ford home. "It sounds like one. Did you give them your credit card
number?"

"Mom, they're not *asking* me for money. They're *giving* me money,
and I've got to move to Ireland to get it. Next week."

Just as the three were getting ready for their big Irish debut, Datar decided to make some helpful introductions, including to Andras Forgacs, CEO of Modern Meadow, and by then a hero in the cultured-animal-product world, having recently pulled in $10 million of investment from Li Ka-shing's Horizons Ventures. Forgacs suggested an alternative. Rather than waste his time in Ireland with what he called a dysfunctional program, why not come to New York and interview with Modern Meadow?

Pandya, who'd now been named CEO of the company, could hardly believe it. His dream of working in the cultured-animal-product space was coming true, and he was only in his early twenties. "I love Andras and respect him so much, and here he was offering me a chance to help culture leather for years to come! And instead I was thinking about starting a company with people I barely even knew at a program that couldn't accelerate itself, let alone a groundbreaking start-up. I've got to be honest: working for Andras was the safe bet."

The three debated it, with Gandhi even conceding that were he in Pandya's shoes, he'd take the Modern Meadow job. Things weren't looking good for the motley crew. But after sleeping on it, Datar suggested that they discuss it with another Horizons-backed company, Hampton Creek.

Josh Balk stepped in. "Look, Ryan," Balk began via a Skype video chat from the Humane Society headquarters near Washington, DC. "Who cares if you're young? You have a cool idea. No one's going to do it if you don't. You're already over there. You're at bat. Are you really not gonna take the swing? Yeah, you might strike out, but what if you hit a home run? You'll just never know if you don't step up to the plate and try."

And that was it. The new milk company was going to become a reality. A few weeks later, the team finally met in person for the first time.

With the company now founded, Gandhi, Pandya, and Datar began figuring out how to actually make yeast cells bend to their will and create milk proteins. Day and night they experimented, manipulating the cells, feeding them different nutrients, keeping them at different temperatures. As they came closer to their goal—but still nowhere near perfection—New Harvest wrote about their work in its newsletter, catching the eye of an editor at *New Scientist*, who was impressed enough to reach out and offer the chance to publish a column in the magazine about their work.

"It was a no-brainer," Pandya recalls. "They were going to let us write an article in our own words, about our own company—and they were going to pay us for it!"

So it came to be that, in June 2014, Pandya published his piece under the *New Scientist* "Big Idea" section, headlined: "Don't Have a Cow: Making Milk Without the Moo." The subtitle was perhaps even more provocative: "If we could make milk the same way we make beer, the environmental benefits would be huge."

In the piece, Pandya wrote that there are four proteins in the casein family, and primarily just two in the whey family, and that those six combined with some water, sugar, and minerals form the essentials of cow's milk. Making it, Pandya explained, didn't involve much new science at all. In effect, the crucial cow's milk proteins have amino acid sequences that anyone can look up online for free. You just convert that sequence into a DNA sequence and order the DNA from any one of many research companies that service the medical sector. That's when the genetic engineering comes in: with the DNA in hand, you insert it into yeast, using a chemical or electrical stimulus. From there, the yeast does all the work, pumping out the proteins you coded it to, similar to how brewer's yeast pumps out alcohol.

Interestingly, even though typical beer production doesn't rely on genetically engineered yeast in the way Perfect Day's milk does, in both Canada and the United States, winemakers are approved to use a genetically engineered yeast that helps improve taste and prevents the creation of histamines in the wine, which can give some people headaches. Another genetically engineered yeast used in winemaking reduces the presence of a carcinogenic compound (ethyl carbamate) that's created during the wine fermentation process.

So Perfect Day is creating the key milk proteins from yeast, but milk is much more than protein. It also has fat, for example, but conventional milk processors already extract the fats from milk and then re-add them in to get a consistent, homogenous product. That's what Perfect Day is doing, but with plant-based fats that are healthier than those from animals.

The thought of scientists the world over reading about their idea was enough to keep the company cofounders' endorphins flowing, but one reader in particular was moved by the piece.

Solina Chau, the face of Li Ka-shing's Horizons Ventures, took interest in the start-up after reading the story. Already a backer of Modern Meadow and Hampton Creek, Chau is among the most powerful women on the planet, according to *Forbes*. She invests primarily in tech ventures she believes are both financially promising and offer the potential of doing good in the world.

Chau first reached out to Balk, who was now serving on retainer as an adviser to Horizons Ventures for its food-tech investments. Needless to say, Balk vouched for the new company and suggested she connect with them. Soon after, Chau asked Frances Kang, director of Horizons and another woman who'd climbed to the elite levels of the venture capital world, to get in touch with Pandya.

As advanced as the bioengineering technology Perfect Day was using was, more basic technology failed them at this key moment. Kang emailed the company asking to meet with them about potential investment, only to have the message languish in their spam filter. She then sent a private Facebook message with similar results, followed by a LinkedIn message. Only silence was returned, and Kang was ready to give up.

But as fate would have it, Pandya—ten days after it was first sent—did check Facebook's "Filtered" folder and saw Kang's unread message. "Oh my god!" the CEO shouted, simultaneously excited by Horizons' interest and horrified that they hadn't replied for so long. "Horizons wants a pitch deck from us!"

Unfortunately for them, the young start-up with its accelerator cash had no pitch deck to share. They got to work on one and quickly sent it to Hampton Creek CEO Josh Tetrick for feedback.

"It's shit," Tetrick replied. Tetrick knew about working with Horizons, and knew Perfect Day's pitch wasn't going to cut it. "Look," he told them from his open office space in San Francisco, "your next call is on Friday, right? You have seventy-two hours to make something better."

Tetrick suggested they go onto a tech freelance site and ask vendors to bid on their project. "If I paid five-thousand dollars to make a pitch deck and it got me a million bucks from Solina Chau," the CEO said, "I'd consider it money well spent."

Pandya and Gandhi took his advice, spending $5,000 to work with a consultant on the proposal. A few days later, they had their pitch deck and it was in Kang's hands. To their relief, she told them she loved it and wanted to meet with them and Balk in Hong Kong the following week. Kang even asked them to bring a sample of their milk.

Gandhi, Pandya, and Datar couldn't believe what was happening, but the next week was particularly bad timing. Experiments they were running needed them there, and they had personal obligations, too. They weren't sure whether to ask for a more convenient date, so they turned to Balk to ask what he thought they should do.

"Let me be clear," the Horizons adviser began. "If my wedding was next week and Solina Chau asked me to meet her in Hong Kong then, I'd postpone my wedding and go to Hong Kong."

―――――――――――

"We had to bring our milk from Ireland to China, but we had no idea how it would hold up," Gandhi jokes. "We didn't think it needed refrigeration, but who knew? This was too important of a meeting to risk it." Lacking better equipment, the duo used a metal water bath to heat water up to the pasteurization temperature of milk, at least 161 degrees Fahrenheit. Less than an hour before their flight to Hong Kong, they finished formulating their samples and held them in the makeshift pasteurizer for as long as they could.

Finally, the milk made its way into a plastic bottle with a screw top, and then into a box of dry ice. Pandya insisted on carrying it by hand the whole way, refusing even to put the box in a check-on bag. "These folks had put tens of millions into companies like Hampton Creek and Modern Meadow. What if we stood a chance of that happening to us? That box wasn't leaving my sight."

Just one problem: security agents in London's Heathrow Airport. "A couple of young brown dudes carrying a box with tiny laboratory bottles of milk? I don't think so," Pandya laughs. "We so clearly did not have kids, and this was so clearly not the way people carry milk."

While Datar was flying in from Canada, Pandya and Gandhi were now held up for more than an hour at security as they tried to explain what the bottle contained. Finally, Heathrow's security detail decided that no matter how dangerous this milk was, it couldn't do any damage from the cargo hold, so the founders were permitted to check their samples, but only after waiting two more hours for the airline desk to open. When they got to the gate, the airline was so apologetic for the hassle they'd endured that they upgraded Pandya to business class for free. (Sadly for him, Gandhi was left behind in coach.)

Now in Hong Kong, Balk, Datar, Pandya, and Gandhi huddled together at the hotel where Horizons was putting them up. It was Balk's first time meeting the latter two, and he knew they must be nervous. "I'll tell you something: just a few years ago I was here to pitch them on Hampton Creek. I know how you feel," he counseled them, recalling how important Chau's seven-figure investment in Hampton Creek was. "They may have more money than anyone you've ever met, but they're still just everyday, friendly people. And you should treat them like you'd treat anyone: be honest, and of course, make the argument why investing in your company would be a smart decision."

"Sounds good," the founders said, poorly masking their nervousness.

"And one other thing"—Balk raised his index finger as their huddle was breaking—"don't kill Solina. If your milk kills Solina, things will go really badly for you."

The next day, Horizons treated the North Americans to a light, casual lunch, before getting down to business. The milk was now being chilled in a fridge. There was no formal presentation, but Gandhi and Pandya used a whiteboard to sketch out various scientific formulas to explain their process to Chau and Kang. Chau was especially engaged, asking detailed questions indicating she understood exactly what they were talking about. Balk, on the other hand, with his degree in political science, merely observed, trying to avoid showing how in the dark he was. "They could've been putting Egyptian hieroglyphics up there for all I knew."

After several minutes of questions and answers, Chau wasn't wasting time. "All right, let's get the big reveal—time to see how it tastes!"

Anxious, Pandya brought out the plastic bottle. "I was surprised he didn't put it in some type of fancy carrying case or another less humble container." Balk grins, shaking his head.

Pandya poured their creation—barely enough for everyone to have a few sips—into coffee cups. As they each raised their portion into the air to toast, Chau intervened, quipping that since Balk had vouched for the company to her, he should be the first one to taste the milk. "It was a bit like being a taster in the royal court," Balk recalls.

His first reaction, as the milk passed his lips, wasn't even how the product tasted (the longtime vegan wasn't exactly a good judge of whether it was like cow's milk or not anyway), but rather to think about what his most effective response to it would be. "I didn't want to oversell it and have it be a letdown when they drank it, but I also didn't want to bias them against it by saying it sucked. In reality, it was somewhere in between."

Then came the moment that mattered. Chau raised her cup, drew a sip, and before the venture capital titan could even swallow, wrinkled her face in disgust. "This is terrible," she stated matter-of-factly.

Perhaps Perfect Day's product was not, indeed, perfect, but it turns out that Chinese milk drinkers are accustomed to ultra-high-temperature pasteurized milk—a sweeter beverage with a caramel flavor that American milk drinkers aren't used to. Pandya and Gandhi had, understandably, formulated their product for the American palate. "We could easily add that flavor once we start selling in China," Gandhi quickly assured Chau and Kang.

And with that, Horizons wanted to sign a deal. "I couldn't believe it." Pandya shakes his head. "Right after she told us she hated it, we were talking about how much they'd invest and what percent of the company they wanted. I'd never seen, let alone even envisioned, anything like it."

By the end of the meeting, the $30,000 they'd been elated to get just weeks earlier seemed like pennies. Horizons had agreed to invest $2 million.

Just a couple of months prior, these three young idealists had never met. Through a series of online chats, they'd come up with an idea for a company to disrupt the dairy industry that brought them from America to Ireland, and finally to China. And sitting at the Hong Kong airport headed home after their successful trip, they were now a seven-figure company.

———————

Two years later, in 2016, now with its own facility in the Bay Area and a dozen employees, including seasoned scientists from brands like Chobani and Beyond Meat, Perfect Day is making milk. With four clear glass fermenters hooked up to a computer—what they call their four udders and brain—their yeast churns away in a milky liquid.

"It takes us seventy-two hours to go from yeast to milk." Gandhi beams. "For a cow, it's two to three years before she can produce any milk at all. And you need all the resources to produce the hooves, intestines, horns, eyes, and all the other parts we don't care about. At Perfect Day, we just need an udder and a brain."

These "udders" are actually fermenters known as bioreactors—also used to make multivitamins, rennet for cheese, and tons of other everyday products. But Perfect Day hopes that, rather than having an udder that can hold a couple of liters of milk, they can have an "udder" that's literally the size of an office building. From there, they can really start competing with factory-farmed dairy.

The yeast cells right now dine on sugars that help them pump out milk proteins. But as Perfect Day continues refining the process, they think they'll one day be able to use grass to feed the cells instead. "Wouldn't that be cool?" Pandya ponders. "To produce 'grass-fed' animal-free milk?" Pandya isn't sure if this would produce a superior product or not, but it may be yet another environmental advantage for the company.

In late 2016, in the hopes of attracting support from the environmental community, Perfect Day funded a life cycle analysis of its milk production as compared to conventional dairy. The results were dramatic. Because they're producing just the milk proteins and not the rest of the cow, the analysis—which isn't peer-reviewed—showed that their milk involves between 24 and 84 percent lower energy use, 98 percent lower water use, 77 to 91 percent lower land use, and 35 to 65 percent lower greenhouse gas emissions.

The company hopes to produce not only fluid milk but milk products as well—like yogurt. Examining what various yogurts look like on a rheological machine (an instrument used to study the flow of liquid matter) in the company lab, it's easy to see the difference between high-end and low-end yogurts. It's not that the milk comes from different cows but rather how the milk is processed after the cows are no longer in the picture. Perfect Day's yogurt still isn't exactly identical to cow's milk yogurt, but it's simple to see on the screen just how much closer it is than yogurt made from plant-based products, like almonds, soy, or coconuts.

The start-up cofounders offer me a chance to compare a plant-based yogurt and their product. As they're preparing the samples, in walks Mark Post.

"I heard there's a tasting of some in vitro yogurt?" the doctor jokes, in town for the first-ever conference on cellular agriculture, which New Harvest had hosted the day prior.

As Post and I put the different products on our respective spoons, it was immediately clear, just by looking at them, which was soy-based and which was cultured cow's milk-based. And once we tasted them, well, let's just say the task became even easier. Whereas the plant-based yogurt was very sugary—perhaps to mask the otherwise less appealing taste—the cultured product was creamier, smoother, and thicker. I don't know how it compared to conventional yogurt, since I haven't eaten it in so long, though my guess is that given how early they were in the process of producing yogurt, Perfect Day probably had some way to go. But they were making progress.

Post perhaps put it best: "The people who want to keep producing dairy cows won't like you, but the planet needs your product on the market."

While Pandya was still a student in Massachusetts, he didn't know it, but another Bay State undergrad, Arturo Elizondo, was also coming to terms with the suffering of animals used in food production. During his internship at the White House during the first term of the

Obama administration, the Harvard government major came across an article forecasting an explosion in China's meat consumption as the nation pulled itself out of poverty. Elizondo was thrilled that more Chinese citizens were moving into the middle class, but was devastated to learn about the environmental implications of such a huge number of people upping their meat intake. At that time, the Obama administration was feverishly working to require fuel efficiency improvements for automakers, yet Elizondo kept wondering, "Why was no one talking about the elephant in the room? If we didn't find environmentally friendlier ways to satiate the coming boom in demand for animal products, whether cars are at twenty miles per gallon or thirty is gonna make little difference."

Later, working in Geneva on a paper for the United Nations Food and Agriculture Organization about food security and shortages, Elizondo came across some writings by Jason Matheny and, to use his own words, "became obsessed." He knew curbing consumer habits is so difficult, especially for animal products. "We needed to provide an alternative, and reading Matheny made me think of this type of technology as a silver bullet. It's the same product, but without all the problems. I wanted to know everything I could about what was happening in the cultured-meat space."

As with so many others, Elizondo's interest eventually led him to contact Matheny directly, which he did in 2012, a couple of years before Pandya and Gandhi. At the time, Matheny was still running the then-all-volunteer organization, and he agreed to Skype with Elizondo. "That one conversation had such a huge impact on me," Elizondo recalls. "It was because of Jason that I signed up for a philosophy class the next semester and began wondering if my dream of becoming an ambassador would really be the best use of my time."

Now voraciously seeking out writings on utilitarianism and effective altruism—to determine the most impactful career path he could choose—Elizondo was introduced to the work of bioethicist and philosopher Peter Singer, including his seminal book, *Animal Liberation*. "That was it: I was a vegan overnight," Elizondo says. "We've been taught that all these characteristics about humans make us special: tool use, language, opposable thumbs, and so on. Yet it turns out that lots of other animals have those same characteristics!"

This change in Elizondo's worldview brought him into contact with plant-based start-ups, like Hampton Creek, along with the cultured pioneers at Modern Meadow. He quickly consumed each article he could find about them, even setting up Google News alerts so he could stay up to date on what was happening. "I just loved the idea that you could have food that provides an identical—or even better— experience for the eater, but that causes so much less harm to animals and the planet," he says. "It was the power of business to do good in the world, and I was so ready to be a part of it."

As he was finishing college, Elizondo had a choice to make. With a Harvard degree, and internships at the White House and the Supreme Court on his résumé, he knew he could take his pick of government service jobs. But a friend and mentor from Harvard, Ben Happ, kept pushing him toward food innovations instead. "Arturo, you're obsessed with food tech now," he recalls Happ saying. "Don't you think you could do more good in the world by helping solve our biggest environmental problem than you could as an ambassador?"

The argument made sense, but he figured that without some type of science or engineering background, what difference could he actually make in that space? "What about venture capital?" Happ pressed his friend. "These companies need money more than they need your expertise."

That was all the pressing Elizondo needed. He booked a one-way ticket, packed his bags, and moved to the Bay Area—with no place to live—hoping to land a job at a major venture capital firm.

While no venture capital company seemed interested, a hedge fund did offer him a job—which was enticing notably because of the six-figure salary. "I thought about taking it just so I could make a lot of money and donate it to animal and environmental causes, but I wasn't sure what the best thing would be."

Like several other young food idealists contemplating what path to take with their lives, Elizondo reached out to Josh Balk. "I was really more interested in seeing if there might be a business role for me at Hampton Creek than anything else, to be honest. But I also just wanted Josh's opinion of what might be an effective path for me to choose."

Balk was glad to talk with him, and suggested they attend an upcoming Bay Area food-tech conference together. As fate would have it, though, the day Balk and Elizondo were supposed to attend the conference, Unilever filed its suit against Hampton Creek over the labeling of its mayo and Elizondo had to attend the conference alone.

As soon as he arrived, he started scanning the room, noticing that just about everyone appeared to be decades his senior. They also all seemed to know one another. Luckily, at one table sat two fellow millennials, one of whom was Datar, by now the new CEO of New Harvest and cofounder of Perfect Day, and Datar's friend David Anchel, a PhD molecular biologist.

His fellow twentysomethings shared his passion for competing against factory farming with food technology, and he learned that Anchel was thinking about trying to culture eggs from a cell taken from a hen's oviduct (where eggs are transported from an ovary). But when Datar explained the yeast-based method Perfect Day was pioneering to make milk, Anchel began wondering why the same couldn't be done for egg whites. (Yolks are more complex and have more structure than the liquid whites, meaning they're more difficult to culture from yeast.) Just like Perfect Day found with milk, egg whites are essentially a few simple proteins plus water, and are so fluid that they don't require the same type of complicated tissue engineering that culturing meat does.

A week after the conference, Datar called Elizondo to inform him that there was an IndieBio accelerator event coming up, the same type of accelerator that had given Perfect Day its first $30,000. "What if," Datar asked, "we started a company to make egg whites from yeast?" Elizondo loved the idea, but wondered what his role would be. "David could handle the science," Datar continued. "I'll help us get funded and make all the right introductions to investors and other VIPs, and you could handle the business side as our CEO. What do you think?"

The CEO? He'd come to the Bay Area hoping to get a job at a venture capital firm. When that didn't pan out, he was looking for some type of work at Hampton Creek to help replace eggs with plants. Yet here he was, at twenty-three, being given the chance to help found and become the CEO of his own cultured-egg company. He didn't need time to deliberate on the offer. "Hell yes!" he shouted into his smartphone. "I'm in."

"I've never been so personally invested in work," Elizondo explains. "It's like there's a fire burning inside of me. I was so excited about the original pitch—I was researching the market for egg whites and found myself reading articles for fun at three a.m."

The three knew what they wanted to do. Basically, they figured that with yeast, water, and sugar, they could use a bioreactor (or as they and Perfect Day preferred to call it, a fermenter) to pump out the proteins in egg whites that give them their taste and texture. Even taking a small fraction of the liquid egg white market from the egg industry would bring enormous returns. A multibillion-dollar market in the United States alone, egg whites are used in protein powders and bars, sold in cartons in the egg aisle for cholesterol-conscious consumers, and much more.

The new company could hardly have picked a more important market to disrupt from an animal welfare perspective, since the chickens who lay eggs for the egg products market—like liquid egg whites—are often the worst treated. The birds producing eggs for the shell egg market (those sold in cartons in grocery stores) are usually kept in cages that are stocked wing-to-wing at only sixty-seven square inches per bird. (To get a sense of how little space this is, consider that a standard sheet of paper is ninety-three square inches.) As overcrowded as that is, at least that sixty-seven-square-inch standard is the result of guidelines that most supermarkets require their egg suppliers to follow so they can claim their eggs meet an industry certification ("United Egg Producers Certified"), anemic though that certification is.

But egg producers supplying the liquid egg market often don't even have that restriction, since most consumers aren't thinking about eggs, let alone hen welfare, when buying products with eggs as an ingredient, like cookies, cakes, or protein bars. As a result, the birds laying the eggs that get used in the liquid egg product market are sometimes confined even more tightly, at only forty-eight square inches per bird, not even enough space for them all to be fully on the cage floor at once.

With their mission clear, the cofounders drafted a pitch to present to IndieBio. Now they just needed a name. Inspired by the natural-sounding Hampton Creek, a few of the options they brainstormed included Oakridge Farms, Oak Farms, River Valley, and more. But in the end, they settled on Clara Foods, a name that had a doubly important meaning for them. *Clara* means egg white in Spanish—important to Elizondo, who'd grown up in both Mexico and Texas—and was also the name of Anchel's beloved late dog. Since Hampton Creek was named after Balk's late dog Hampton, the group figured it might be good luck. In fact, an early iteration of Clara Foods' logo even included a dog. "We had to get rid of her from the logo, though," Elizondo protests. "It'll be tough enough to overcome the obvious hurdles we'll face. The last thing we want people to think is that we're selling pet food!"

The final logo is a bright yellow sun drenching a green field with its rays, but the sun is actually an egg yolk, and its rays part of the egg white. The font has a natural tone; one that feels very familiar. It should be familiar, since it's purposefully the same font as Whole Foods Market's logo, except that unlike the natural foods giant, the text isn't in all caps.

Elizondo explained Clara Foods' mission at the IndioBio accelerator: "More than anything we want to provide an alternative to people who want to consume products that align with their values, their environmental conscience, and, ultimately, products that are a good source of protein without all the baggage."

In addition to "no baggage," just as Perfect Day is making milk without cholesterol or lactose, Clara Foods wants to take advantage of the versatility of their yeast-based process by making products that are better than eggs from chickens. "More protein? Easy. No *Salmonella*? Obviously!" Elizondo laughs.

Part of the pitch to early investors included just why the company thought it would be so efficient compared to conventional egg production. After all, think about what it takes to produce an egg. You need a chick (and the whole system in place to produce those chicks), and then you need to feed and house her for four months before she even starts laying eggs. You need housing, lighting, climate control, feed, water, supplements, medications, labor, and much more. Even once she starts laying, she's only laying on average one egg per day. With fermentation, of course there are still resources needed, but Clara Foods argued it would soon be able to pump out a constant flow of liquid egg whites for consumers with a small fraction of the inputs its competitors in the egg industry needed.

The pitch worked. As it had with Perfect Day, IndieBio agreed to fund Clara Foods, granting them office and lab space, as well as $50,000 to get started. And with that, the team set to work in December 2014.

"We figured all they'd need to do is take five or six proteins, made from yeast—just like beer—and put them together and we should get an egg white," recalls Ryan Bethencourt, the SOSV venture capitalist behind IndieBio and an early investor in Memphis Meats. "We didn't know what might happen, but we figured it should work. We're only interested in results, not just science for science's sake. We wanted an egg white."

Clara Foods' first task would be to provide some type of a proof of concept. The goal: take an actual egg white, reduce it to its protein components, and then reassemble those components into an egg white that cooks and tastes just like a typical egg white. Within months, using the same type of fermenters that serve as udders in Perfect Day's nearby lab, they'd done just that. If they could build an egg white from isolated egg proteins, now all they needed to do was figure out how to make those proteins with yeast.

As word of their efforts spread, Elizondo knew it was only a matter of time before big investors would want a piece of the action. The market for egg whites is gigantic, and he knew Clara Foods stood a high chance of being able to produce them at a more affordable price than the conventional egg industry could—and without the food-safety risks.

And he was right.

Just four months after Clara Foods launched its lab, a group of investors, again including SOSV, joined together to pump $1.7 million into the start-up, launching it into the public eye and earning the attention of the broader tech industry. Since that infusion of cash, the company has taken off, hiring more than a dozen scientists and planning its first commercialized product—probably an egg white protein supplement that could be mixed into the bars athletes use—to hit the market in early 2019.

———————

What Perfect Day and Clara Foods are doing—essentially reverse engineering cow's milk and egg whites by building them from the protein molecule up—offers a whole host of new ways to create food products with greater functionality than any that have come before. As long as you know what key proteins make up the product you want to create, you should be able to get some yeast or other microorganism to generate those specific proteins for you. Such technology paves the way for products few people would've ever imagined, and one cellular-ag start-up is taking these possibilities to (pre)historic heights.

As we saw with Modern Meadow, collagen is really the building block of our bodies. It's by far the most abundant protein found in animals, and because of its ubiquity, we've found all kinds of uses for it. It's of course essential to leather, but it's also important for gelatin, the flavorless extract from animal skin and bones used in a variety of products, from Jell-O to cosmetics. We may have a more granular understanding of the collagen molecule than any other protein, simply due to decades of study.

"With so many cellular-ag foods, you're worrying about the flavor, or the smell, or something else," Alex Lorestani, the CEO of Geltor, the start-up culturing its own gelatin, says. "With gelatin, all you care about is its stiffness." In other words, if you're making Jell-O, you want what's called "low bloom," or relatively soft gelatin. For gummy bears, you need high-bloom gelatin to make the candy firmer and chewier. The quality is referred to as "bloom" because the test to determine how stiff gelatin is was invented by Oscar Bloom in 1925.

Since gelatin is basically just an extract of collagen, replicating it from scratch by getting microorganisms to produce it shouldn't be that hard, Lorestani and his grad school classmate Nick Ouzounov figured in 2015. The two microbiologists agreed that they ought to be able to produce gelatin without animals by using a microbial production platform, starting with bacteria, just as Perfect Day and Clara Foods do with yeast. (Recall that human insulin and rennet for cheese are also produced by bacteria, not yeast.)

Similar to meat, milk, and eggs, there are already plant-based substitutes, such as agar and carrageenan, for gelatin on the market today. But while those products have some uses, according to Lorestani they're about as good a replacement for animal gelatin as the first vegetable patty was a replacement for an actual hamburger. "I'm crazy for plant-based proteins," the thirty-year-old CEO with a shaved head offers, "but current substitutes for gelatin fall far short for most gelatin uses. Have you had an agar gummy bear? They sometimes break in your teeth; it's not good."

The two scientists joined forces to propose an idea to IndieBio for a start-up that would create gelatin in fermenters. Their idea struck Bethencourt's fancy, and in August 2015, Geltor was granted some new cash and a lab space that, at the time, they shared with Clara Foods. Elated to have a new venture, Lorestani and Ouzounov knew they needed something to break out, and it had to be big. Really big.

When humans arrived in North America some eleven thousand years ago, they found a continent filled with gargantuan animals. Mastodons may have been the biggest, but sadly for these long-tusked relatives of Asian elephants, they didn't evolve around *Homo sapiens* and consequently were little match for us. Very quickly, they and other so-called megafauna found themselves endangered and then totally extinct. Some of the vanquished beasts remain today, however, sealed shut in icy graves that have preserved their bodies for millennia. And as with all ancient organisms, if there's any protein still to be found on them, it's probably going to be in the form of collagen. Indeed, humanity has now taken in some ways an initial step toward resurrecting the mastodon, at least at the molecular level, by sequencing the long-gone animal's proteins. Anyone with an Internet connection can freely access the mastodon's protein sequences in just a few seconds.

Knowing this, in late 2015, just like Perfect Day was ordering DNA for milk proteins, Geltor sent an order to a DNA printing company to get a vial of DNA-encoding mastodon collagen. Once secured, the scientists put it through their process and started producing actual mastodon gelatin. Lorestani and Ouzounov could've made gummy bears, but the two cofounders thought it'd be cooler to order an elephant mold from Etsy instead. (They couldn't find mastodon molds, but it'd be hard to tell the difference anyway, they figured. Gummy elephants would have to do.) Very quickly, after mixing their gelatin with some sugar and pectin, the world had its first mastodon gelatin candy. Watching Ouzounov put the small gummy elephant in his mouth, Lorestani thought, "Man, this is the first time anyone's eaten mastodon protein in a really long time." Talk about a Paleo diet.

When asked how it tasted, Ouzounov explained that he was really more concerned about optimizing for texture than taste at that point. Perhaps adding any type of flavoring would've helped (they omitted flavor from the first batch so as not to distract the palate). The good news is that Geltor isn't interested in selling mastodon-based gummies directly to consumers but rather intends to produce ingredients for food manufacturers who need gelatin, and not necessarily of mastodon origin, as the company is now producing gelatin from more conventional animal DNA. In 2016, the company raised well over $2 million in venture capital funding, primarily from SOSV, New Crop Capital, Jeremy Coller, Stray Dog Capital, and legendary biotech venture capitalist Tom Baruch. Now with a dozen employees, Geltor had one primary mission: to become the first of the new group of acellular-ag companies to commercialize its product.

And they did.

The company's first products hit the market in mid-2017, with some of its initial customers being cosmetics companies eager to replace conventionally sourced animal gelatin with Geltor's cleaner, more functional, and animal-free version. Medical testing labs that like to experiment on collagen are also placing orders, hoping that the new product will have better properties for their experiments, too.

"As a food community, we're settling for protein production platforms we have at our fingertips," Lorestani observes, sitting in his Bay Area office—shared with Memphis Meats—cloaked in his signature gray hoodie. "Mostly that means exploiting animals, or in some cases plants that are really abundant. We're really good at producing huge numbers of animals, and it's worked okay for a while. But today animal agriculture is a big strain on our civilization, and we can do better. That's what we want to show."

———————

Perhaps few people's mouths water when they think about eating milk, eggs, or gelatin derived from genetically engineered microorganisms, whether yeast or bacteria. Then again, the conventional methods used to make these products hardly make for good dinner table conversation, either. In the case of gelatin, how many people really want to eat hydrolyzed collagen from an animal's skin and bones that has marinated in an acid bath for a month? Or milk from a cow who was pumped full of hormones and antibiotics? Or eggs from a bird confined in a cage so small she never spread her wings?

With Perfect Day, Clara Foods, and Geltor now funded and readying themselves for market penetration, they still face the same pressing question that's troubled those in the clean-meat community: how to get consumers on board. Yes, supporters often refer to meat, milk, and egg "breweries" as a way to describe the process that goes on inside these labs in terms that laypeople can understand. But there are some key differences between beer brewing and what these start-ups are doing. First, humans have been consuming beer for millennia and are familiar with its basic functionality. Second, beer doesn't depend on twenty-first-century genetic engineering, as the yeast and bacteria used by companies like Perfect Day, Clara Foods, and Geltor do. While these genetically modified microorganisms (or "designer yeast" and "designer bacteria" as supporters call them) don't make it into the final products, rendering them GMO-free, the fact that they're used in the production of these foods is still enough to give some GMO antagonists pause.

Perhaps because they're so much closer to market than meat, or perhaps because they involve using genetically engineered yeasts or bacteria to produce (unlike the meat-makers, who are using tissue engineering but not genetic engineering), the opponents of biotech tend to be more focused on criticizing these acellular-ag products, just as they are with some other similar products already in use.

Take, for example, vanilla.

There are a multitude of concerns about vanilla production, most notably that it must be produced in the rainforest, where the vanilla orchid grows. There's nowhere near the capacity to grow the amount of vanilla that the world demands, making it among the most expensive spices on the planet. The good news for vanilla-lovers is that it turns out that vanillin is the compound that makes vanilla smell and taste like vanilla. And we've long known how to produce vanillin without farming the rainforest, enabling a more reliable and much more economical version of the flavoring so many of us want. (Natural vanilla extract costs many times more than synthetic vanillin. On a per ounce basis, it's comparable to truffles and saffron.) For these reasons, nearly all the vanilla flavoring used in food today no longer comes from the plant but is instead a synthesized version, typically produced from petrochemicals or wood pulp, which raises its own sustainability concerns.

But one Swiss company, Evolva, has found a way to brew its vanillin simply by fermenting yeast and making it produce the coveted flavoring all on its own. To many, it's a sustainability success story. To others, like Friends of the Earth, it's "an extreme form of genetic engineering" that should be shunned in favor of the more natural rainforest-harvested vanilla that currently comprises about 1 percent of the market.

The reason for the concern: Evolva gets its yeast to produce vanillin by editing its DNA, leading the tiny organisms to produce vanillin that's identical to the kind we eat today. And that process is exactly what newer acellular-ag companies are replicating. Not all environmentalists are with them, however.

"These synthetic biology techniques present even more concerns than the first generation of GMOs," Dana Perls of Friends of the Earth argues. "Like conventional GMOs, synthetic biology products are starting to enter the market with virtually no health or environmental assessment, oversight or labeling."

By "synthetic biology," Perls is referring to the new branch of science that applies engineering principles to biological processes. Whereas conventional genetic modification of plants and animals (aka GMOs) involves splicing genes from one species into another, or editing genes to "knock out" certain genes in an organism, synthetic biology allows scientists to make new DNA sequences altogether. The benefits include the ability to design all types of new organisms, like yeast and bacteria that can perform fully novel tasks such as producing milk, egg, and collagen proteins, but also medicines, biofuels, and perfumes.

Where Perls sees peril, those in the "synbio" community see promise. The idea of efficiently creating so many of our resources from single-celled organisms as opposed to mining, drilling, or farming them offers a chance to lighten our footprint, they argue. "What is at stake here is finding a way to make everything humans need without trashing our civilization," Drew Endy, a Stanford University synthetic biologist and early synbio pioneer told *Newsweek* in 2017. "We can transition from living on Earth to living with Earth."

Even though there's a difference between GMOs (which are in more than 70 percent of packaged foods sold in the United States) and synthetic biology (which, as of this writing, is used to produce very few foods), both are viewed with great skepticism by groups opposed to biotech in food. A lot of people just instinctually don't want scientists tampering with their meals, even though nearly everything we eat today is the product of science, including fruits and vegetables, which have been genetically selected (though not genetically engineered) to the point that they barely resemble the foods our ancestors ate. For example, if you've never seen what the original North American corn looked like, it's nothing like what you'll find at your local farmers' market today, certified organic or not. Instead, think about something much more similar to a small pinecone than today's giant-kerneled food we need two hands to eat. The same type of artificial selection process occurred with numerous other foods we routinely eat, from bananas to tomatoes, and more, thankfully for us.

While there's a legitimate debate about whether GMOs are better or worse for the planet, there's been little scientific evidence that shows GM foods are any less safe to consume than other foods, but that hasn't stopped consumers from fearing them. In one particularly compelling video from 2014, *Jimmy Kimmel Live* sent an interviewer to a farmers' market to ask shoppers if they avoided GMOs. Invariably the respondents said they did, mostly for concerns about health, but many then admitted that they don't know what the acronym GMO stands for. As one of the interviewees jokes when asked what a GMO is, "I know it's bad, but to be completely honest with you I have no idea."

The battle over GMO labeling alone has cost both sides tens of millions of dollars, leaving a lot of consumers confused yet still wary. That said, there's a key difference between genetic engineering and synthetic biology: GMOs are largely (though not entirely) produced by megacorporations like Dow AgroSciences and Monsanto, in part to maximize output of feed crops for animal agriculture. Synthetic biology for agricultural products, on the other hand, is primarily being used by tiny start-ups seeking to solve key environmental problems by replacing traditional animal ag (including the GMO crops used to feed farm animals, which represent nearly all GMO crops in commercial use) altogether. As author McKay Jenkins points out in chapter 1, little would do more to reduce the number of acres planted with GM crops than replacing animal ag with cellular-ag products, even if some of them use GM technology in their production process.

Recognizing the passion often involved in the battles over GMO labeling, Elizondo pushes back, arguing that consumers often don't understand the basic science behind the food they eat. To illustrate, he points to a 2015 Oklahoma State survey that found that more than 80 percent of Americans, the same number who say they support labeling foods containing GMOs, say they support "mandatory labels on foods containing DNA." As readers of this book likely already know, nearly every food humans have ever eaten (with some exceptions like salt) contains DNA. "And," Elizondo continues, "how many production practices should be labeled? Since nearly all cheese nowadays is made with rennet, which is the result of a similar process using genetic engineering, should we be labeling cheese as GMO, too?"

Michael Hansen of the Consumers Union isn't buying it. A vocal critic of GMO technology, Hansen sees Elizondo's, Pandya's, and Lorestani's products as all peas in the same genetically modified pod. "It doesn't matter that the genetically engineered yeast isn't in the final product," Hansen asserts. "These products are the result of genetic engineering." When asked if he feels the same way about cheese with genetically engineered rennet, he rebuts that rennet is a small component of cheese, whereas these products would not exist at all without genetic engineering. "Most of the cheese is still natural," he claims. "These foods are entirely lab-produced."

This all raises the issue of just how cultured-animal products—if the barriers to entry, such as cost and regulatory hurdles, are overcome—ought to be introduced into the marketplace. If consumer acceptance was really so low, whether based on technophobia or on legitimate concerns about these novel foods, should these start-ups—if they don't have to—disclose that there's anything different about their products?

Consider that while the majority of papayas sold today are genetically modified (they've been engineered to have protection against a common virus that can damage papayas), you don't see the fruit labeled as such. People simply buy papayas, not papayas advertised as GMO (negative connotation), nor those labeled "virus-resistant" (positive connotation). Lorestani is hopeful on this point, again driving home the rennet argument used by the champions of acellular agriculture. "There's really nothing categorically different between the cheese you eat, which has genetically engineered rennet, and a piece of candy you might soon eat, which may have GE-produced gelatin."

The confusion surrounding GMOs is so great that some companies have taken to labeling their products for no reason as a way to distinguish themselves from others. For instance, tomato producer Pomì has begun advertising its tomatoes as "GMO-free," even though there's no GMO tomato on the market. (It's kind of like the trend of advertising bottled water as "gluten-free.")

Like Hansen and Perls, Memphis Meats founder Uma Valeti also wants consumers in the know. And he hopes that providing information about his product will inspire them to choose his meat over its conventional counterpart. "The benefits are so astounding, simply on food safety alone, that informed customers will demand our meat for their family. Who wants to put their kids at greater risk of food-borne illness?"

Elizondo agrees. "Imagine choosing between an egg white that might have salmonella and one you know doesn't. Or milk with pus versus milk without pus, which all cow's milk has. Which would you choose?" (Cow's milk has a certain amount of pus in it, more technically called a somatic cell count.)

The Good Food Institute's Bruce Friedrich touts transparency as one of the key advantages cellular ag has over conventional ag. The more people learn about it, he argues, the more enthusiastic they'll be. "Cultured products are better—they're safer, more sustainable, less polluting, and better for animals. Once they're the same price or even if they're slightly more expensive, telling consumers what they're getting will be a huge selling point."

Perhaps many will choose acellular-ag products, but the optimism of those in the cellular-ag community may be dampened by what a lot of consumers at least say they want. A 2013 *Wall Street Journal* article reported that 51 percent of Americans say they seek out foods labeled "natural," even if there's widespread confusion about what that word actually means. And it's hard to imagine that many people will consider meat, milk, and eggs produced without actual animals to be very "natural."

At the same time, the question of what's "natural" is hardly settled. Are animals who never existed before humans genetically selected them (think of nearly all dog breeds) natural? The same goes for the broiler chickens we eat today who've been genetically selected for rapid growth and obesity. Such a process is hardly "natural," but few consumers seem to take issue with it when it comes time to make purchases in the marketplace.

Skeptics like Hansen think this comparison misses the point. From genetically selecting farm animals for exaggerated production traits (without technically genetically engineering them), to giving animals drugs to make them grow faster, it's very clear to him that our current meat-production system is unnatural, but that's not a reason to move even further away from natural production. Put another way, current systems are unnatural but can be made more natural. Meat without animal slaughter—that, to him, is unnatural *per se.*

Datar sees his point and takes it in the opposite direction. "If we're already so far removed from the fact that a boneless, skinless breast came from a chicken and most people don't have a problem eating that, why not just have it actually not come from a chicken?"

In the end, the natural reaction many have to applying science to food is one companies like those in this book will have to overcome, especially those using genetically engineered microorganisms to produce their foods. It's incumbent upon them to demonstrate that what they're doing isn't different from other ways of producing foods that are already widely consumed (e.g., rennet in cheese); that their foods are perfectly safe and, in fact, are likely safer than the foods they're displacing; and that the environmental and ethical benefits of their production are just too great to pass up.

Some cellular- and acellular-ag companies have already begun trying to help federal regulators understand what they're doing. "We want them involved with us every step of the way," Pandya says. That's why he and Gandhi met with the FDA in late 2016 to introduce themselves and answer any questions the agency might have. Chief among their concerns is avoiding the type of standard of identity dispute that Hampton Creek had with the agency over its use of the term "mayo." The other key to their and other start-ups' success will be proving that their product isn't really different from what's currently consumed and is therefore deemed safe.

In fact, the FDA already publishes a list of "Microorganisms & Microbial-Derived Ingredients Used in Food" it classifies as "generally recognized as safe" (GRAS). Some of these are types of baker's yeast, the yeast used in wine referenced earlier, rennet, vitamin D, vitamin B_{12}, and more. And many of those ingredients are produced via the type of yeast or bacterial fermentation that Perfect Day, Clara Foods, and Geltor are using. In good news for these companies, in late 2016, the FDA finalized a rule that allows food companies to determine if their ingredients are GRAS without having to wait, sometimes for years, for the FDA to conduct its own studies.

The new rule was condemned by groups like the Consumers Union, which argued in favor of modifications to the rule that would've required independent experts to conclude if an ingredient is GRAS rather than let the food company conduct its own studies to make that determination on its own. The FDA retains the ability to challenge any ingredient's GRAS designation, but the company itself now makes that determination first.

In Europe, the government exerts much more control over such designations. The European Commission (EC) publishes a list of "Novel Foods," which it defines as "food that has not been consumed to a significant degree by humans in the EU prior to 1997," the year in which this regulation took effect. Some of these foods, such as chia seeds and agave syrup, aren't actually new but were simply not consumed in the EU until recently. Others, meanwhile, are actual products of biotech, such as oils enriched with phytosterols to cut cholesterol.

Of the ten categories the EC uses to classify these novel foods, one seems specifically intended to address the issue at hand: "Food consisting of, isolated from, or produced from cell culture or tissue culture derived from animals, plants, microorganisms, fungi, or algae." Another of the categories includes, "Food consisting of, isolated from, or produced from microorganisms, fungi, or algae."

Novel foods can be sold in the European Union—many Europeans are happily enjoying their chia seeds today, for example—but only after the EC has determined them to be safe for consumers. Even then, they must be "properly labeled to not mislead consumers."

These are all hurdles the cultured companies will have to clear before they can sell their products commercially. Presuming federal regulators allow these products on the market, it'll be up to companies and nonprofits like those in this book to sell customers and major food brands on their claim that their products are either no different from or better than the animal products we buy today. Given the pace of these companies' progress, it won't be long until we'll find out if consumers will come to accept these seemingly novel foods into their diets or if the investors pumping millions into these startups will wind up with a lot of (cultured) egg on their faces.

8

TASTING THE FUTURE

Just over two hundred years ago, English scholar Thomas Malthus predicted that, because population grows exponentially while food production only increases linearly, unless humanity limited our growth (either by reducing the birth rate or dying off through war and disease), there would inevitably be a collision between the number of humans who inhabit the planet and the amount of food available to feed them all.

Fortunately for us, so far he hasn't been proven right. Malthus didn't foresee the tremendous gains in agricultural productivity that would be made in the twentieth century especially. Excepting the brief anomaly of the bubonic plague in the fourteenth century, humanity's numbers have steadily risen every century for the past couple millennia and began skyrocketing after 1900. We entered the twentieth century with one and a half billion of us but ended it with more than six billion. Today nearly eight billion of us are spread across the planet, and by 2050, on our present course, we're likely to reach nine to ten billion.

In other words, we've had no problem fulfilling the biblical ex-hortation to be fruitful and multiply. But will Malthus eventually be vindicated?

As discussed in chapter two, Norman Borlaug prophesized dur-ing his 1970 Nobel Peace Prize acceptance speech that the green rev-olution he helped usher in bought us some time; it kept large parts of our rapidly growing population in the twentieth century fed. But it didn't offer the promise of a permanent solution. In the absence of a check on the "population monster," as he called it, *Homo sapiens* sim-ply won't be able to keep up with the increasing demand for food that the coming billions of new humans will require, especially as long as we're relying on a protein production system as inefficient as animal agriculture.

Voluntarily reducing our population levels seems extremely un-likely in the next few decades when change will be most critical. This means that, in order to avert a Malthusian catastrophe, we're going to need another green revolution. One such option is simply to eat greener by adopting more plant-based diets, which we already know are healthier, more efficient, and more humane.

Why not shift away from animal-based meat and toward plant-based proteins, since they typically require far fewer resources to pro-duce than animal proteins? As many of the cellular-ag backers in this book attest, they see a bright future for plant-based meats, which is one reason those companies have attracted so much venture capital.

Given these products' pace of improvement, and the increas-ing alarm bells from the public health community about the over-consumption of meat, it does seem likely that plant-based "chicken" nuggets and "pork" sausages will constitute a greater portion of hu-manity's diet in the future. As a consumer, I'm quite content to en-joy these alternatives myself instead of actual animal meat. Perhaps many other people will also be amenable to making them their pri-mary or even sole source of "meat," especially if the companies marketing them can offer their products at more affordable prices than animal meat.

Many clean meat enthusiasts are in agreement on this point: if cellular-ag-produced foods fail because the plant-based protein companies become wildly successful, they'll all be thrilled. They mostly look at clean animal products simply as a concession to human nature—people really want to eat meat from animals, and this is a much better way to produce it.

In my own experience, I've found that many very compassionate and eco-conscious people simply want to eat what they consider "the real thing," either occasionally or all the time. These people love animals, they want to protect the planet, and they care about their health. Yet for a variety of reasons, they just have a hard time making the shift to a plant-based diet, or struggle to stick with it once they start. Even with widespread availability of plant-based meats, the rate of vegetarianism in the United States has remained between 2 to 5 percent for decades. And research by animal advocacy groups themselves shows that 86 percent of vegetarians eventually return to an omnivorous diet.

This isn't to suggest that people won't eat less meat. Even if most don't become full-time vegetarians, there's real evidence that for health reasons alone, many in the United States and the European Union at least are seeking to cut back on the amount of animal products they're eating. This is positive news for the planet, but hardly enough to accomplish what's needed to avert the problems we're already facing, like climate change, environmental degradation, and animal cruelty, all of which are greatly exacerbated by animal agriculture. And it's certainly not sufficient to avert these problems if they continue to get worse in the future.

Presuming the human population continues to grow and most of us continue to want meat, finding a *much* more efficient way to produce real animal products is of critical importance for the planet and all its inhabitants, humans and nonhumans alike. Advancements in cell-culture technologies offer us a chance to do just that, while at the same time addressing other critical problems associated with the factory farming of animals. And, of course, they offer potentially enormous returns for those with the foresight to invest in these start-ups today.

In 2010, with the exception of the synthetic rennet-makers supplying the cheese industry, not a single food company was growing animal products outside of the animal commercially. In fact, not a single one of the companies profiled in this book even existed. By 2020, however, the situation will be dramatically different. In the past few years a handful of pioneering entrepreneurs have invented a whole new field of agriculture, one that could solve many of the most pressing problems we face. In a real way, they have the potential to usher in the second green revolution that's needed to ameliorate the resource crisis Malthus prophesied. Already we're facing tremendous strains and causing mass extinctions, problems we'll face in an even more severe way in the future without intervention now. The clock is ticking, and without finding better ways to feed ourselves, the problems of today may end up seeming quaint compared to what could ensue in the coming decades.

In addition to the future Malthusian dystopia that a clean-meat industry could help avert, there are other more immediate benefits these products could bring. The effects on the meat industry today, for example, could be quite stark.

———————

Some in the conventional meat industry already see the writing on the wall. They may not yet be conceding that we've reached "peak meat," but animal protein companies are beginning to diversify their portfolios. For example, plant-based chicken purveyor Gardein is now owned by Pinnacle Foods—the owner of brands like Hungry-

Man and Van de Kamp's fish—while Kraft Foods owns Boca Burgers in addition to meat companies like Oscar Mayer. And in a bombshell announcement in late 2016, Tyson Foods, the biggest meat producer on earth, purchased a 5 percent stake in the plant-based protein company Beyond Meat. Tyson's then-CEO Donnie Smith tweeted about the announcement, "Excited about our future," with a link to the *New York Times*' coverage of the Beyond Meat investment.

There are some in the industry who are clamoring for more involvement in the clean-meat space, too. In 2016, even prior to Cargill's investment in Memphis Meats, Lisa Keefe, editor of meat-industry trade magazine *Meatingplace*, encouraged her readers to take a look at what Modern Meadow and Mosa Meat are doing, along with plant-based companies like Beyond Meat, Impossible Foods, and Gardein. "Rather than make and distribute only animal-derived, meat-based products, processors might redefine themselves as makers and distributors of protein products—then create, or acquire, companies and brands related to protein, regardless of origin."

Keefe's editorial goes on to admire the capital-raising success of the cultured animal product companies. She wonders why more in the meat industry aren't involved in this type of innovation. "Meat production could do worse than to cozy up to smart investors like . . . the constellation of backers behind Modern Meadow," Keefe wrote. "They see protein production as less of an ag issue and more as a technology issue."

If big meat producers start taking Keefe's advice and begin embracing these culturing technologies, it might help them take steps toward solving a transparency problem they've created for themselves in recent years.

As a result of repeated whistleblowing exposés by animal welfare groups and food safety advocates, the meat industry has been stung time and again by meat recalls, slaughter plant shutdowns, animal cruelty convictions, and more. The industry's response to these incidents largely hasn't been to try to prevent such abuses. Rather, it's been simply to try to prevent the public from finding out about these abuses in the first place, as the *New York Times* headlined in a front-page 2013 story, "Taping of Farm Cruelty Is Becoming the Crime."

In recent years, dozens of "ag-gag" bills have been introduced throughout the nation, all intended to prevent transparency in the meat industry. Some go so far as to make it a crime simply to take a photo or video of a slaughter plant or factory farm. Idaho's and Utah's laws doing just that were struck down as unconstitutional. Others essentially make it illegal for a potential undercover investigator to gain employment at an agribusiness operation. Iowa's law doing this is still in effect.

Frankly, animal agribusinesses have been so successful at thwarting virtually any limits on their conduct toward animals that it's understandable that they'd be so averse to transparency. In the absence of farm animal welfare laws, they've essentially been in a race to the moral bottom, where, in the name of efficiency, inhumane practices have become the norm. For example, were a veterinarian to neuter a dog without pain relief, that vet would likely be charged with criminal animal cruelty. But the same abuse—castration without pain relief—is an everyday occurrence in the pork and beef industries, since farmers have successfully exempted their practices from most state anticruelty laws. And there isn't a single federal law relating to the treatment of animals on farms. Similarly, locking your cat in a cage so small she can barely move an inch for her entire life would land you in jail, but such lifelong immobilization of pigs and chickens is routine in the pork and egg industries.

In other words, it's hard to blame the animal-ag sector for wanting to hide its practices from the public. Were most people to see just how animals are raised for food, they might think twice about whether they really want to eat those animals. Clean-animal-product companies, however, could radically transform the relationship consumers have with their protein providers.

"One of the selling points of clean meat is that, because it's brewed in fermenters, there's total transparency," Bruce Friedrich of the Good Food Institute says. In other words, there's nothing to hide. "Right now, good luck getting into a factory farm or slaughterhouse," he says. "Other than a few free-range operations, the vast majority of farms are not good experiences, and that's true for all slaughterhouses. But you could take a tour of a clean-meat factory, just like you can tour a brewery. I can't wait for Memphis Meats to offer the first public tour of its meat factory when it's ready."

Such a shift toward a more transparent meat industry would mark a drastic departure from its past track record, one that food safety, environmental, and animal advocates would no doubt welcome. The total openness of cellular-ag companies might also go at least some way toward reassuring those concerned about the marriage between food and biotech. Few people are concerned about the use of biotech to produce insulin for diabetics or other lifesaving medicines. But the same exact processes applied to food seem to be more worrying to some at least, and getting to see exactly how these foods are being produced may alleviate some of the fear, uncertainty, and wariness surrounding the technologies.

———

Transparency aside, the future of these new cellular-ag start-ups is still uncertain as they move closer to commercialization and normalization in the marketplace. The questions about biotech raised by critics in this book may also be raised by regulators and consumers alike. But what if these companies succeed? We know it would have a huge impact on the protein-production sector, causing a massive shift in where jobs in the ag economy are located. But it would also come much closer to putting one stakeholder in the food industry out of business: farm animals.

Replacing chickens, turkeys, pigs, fish, and cattle in our agricultural system with our new local neighborhood meat breweries would raise enormous questions about these animals whom we bring into existence to feed us. In short, there'd be vastly fewer farm animals, which of course is largely the point of the new technologies. But not everyone would be so happy with a world with so few farm animals. This line of critique is far more philosophical than practical, and your opinion on the matter really rests on whether you agree with turkey cell culture researcher Marie Gibbons who declared to the *MIT Technology Review*, "It would be better if farm animals didn't have to exist."

It's not that Gibbons dislikes farm animals; quite the contrary. She's a huge animal lover. But Gibbons believes the world would be a better place—for both those farm animals who'd never be born, and for the rest of us—if we had fewer livestock and simply left the land for free-living, wild animals instead.

Not everyone sees it that way. In 2008, the *New York Times* editorial board expressed sympathy for the argument that factory farming must end, but embraced what it called "a more measured approach" than replacing farm animals with petri dishes. The *Times* made the case for reducing animal cruelty in agribusiness but warned that "it will be a barren world if the herds and flocks disappear in favor of meat grown in a laboratory tank."

Some philosophers have taken up the *Times*' argument and have started worrying about what impact a farm-animal-free future would have. They don't favor the current ag system, and in fact, they rail against it. Rhys Southan, a philosopher and clean-meat skeptic, makes his views about factory farming known: "Since the lives of animals who become our food are mostly a curse, producing mindless, unfeeling flesh to replace factory farming is an ethical (as well as literal) no-brainer."

In other words, the animals who are farmed for their flesh, milk, and eggs today are generally so miserable that they'd be better off never being born in the first place. This probably isn't true for beef cattle, who spend a lot of their lives outdoors and with the ability to engage in natural behaviors, but for the factory-farmed chickens, turkeys, fish, and pigs who comprise more than 99 percent of America's farmed animals, it's hard to argue against the point. The day of their slaughter may actually be the best day of their lives, since it finally brings an end to their chronic suffering.

So for these clean-meat skeptics there isn't much of an argument that the current system is morally preferable to a system of breweries pumping out insentient meat, milk, and eggs to satiate our desire for animal products. But what about those farm animals who actually have decent lives? Grass-fed cattle, for example, often live on the range their entire lives, never know a feedlot, and have no cropland devoted to growing corn or soy to feed them. One can argue about the ethics of killing animals for food when it's not necessary (obviously the vast majority of us can survive without eating animals), but would these animals never existing in the first place be better than us bringing them into the world, giving them a good life, and killing them rapidly?

Southan believes not. In an essay entitled, "Execution at Happy Farm," he argues that even with the sad moment of slaughter, a cow who's enjoyed some time on earth and is then quickly killed is better off for being brought into existence. For this reason "kill-free meat might sound nice," Southan warns, "but it's kill-free only because it never had any life to end."

Southan isn't alone in his view. Two scientists at the University of Oxford's Future of Humanity Institute, Anders Sandberg and Ben Levinstein, argue in their essay, "The Moral Limitations of In Vitro Meat," that of course the world would be better with most current farm animals existentially displaced by cultured products. But what if we could improve those animals' lot to make their lives actually worth living? Wouldn't we then prefer that they exist?

Sandberg and Levinstein believe that "if we stop or nearly stop raising livestock, then the sum of pig, cow, and chicken happiness in the world will be approximately zero. Although the sum currently is very likely negative, it would be a shame if virtual extinction of these species is our best moral option."

I'm not sure that "extinction" is the right word for these domesticated animals, as they didn't even exist until a few thousand years ago, or in some cases only a few hundred years. If a breed of dog that didn't exist until we genetically selected it into existence all of a sudden were to stop being bred, would we really consider their absence an "extinction"? The domestication of animals—taking wild animals and selectively breeding them so they're more docile and dependent on humanity for their survival—is a different question altogether, but regardless of what word best characterizes these animals' absence, it's certainly an important moral question.

Sandberg and Levinstein go on to make a case that slashing the number of farm animals who are actually enjoying a decent life would reduce a lot of happiness in the world—happiness that would've been enjoyed by those very animals many clean-meat enthusiasts are seeking to help. The authors concede, though, that those farm animals' existence displaces and harms a large number of wild animals, especially through deforestation to provide pasture or cropland. And as Jason Matheny argues in his 2003 essay, "Least Harm," the wilderness tends to have more animals inhabiting it than pasture, so if our goal were to maximize the number of sentient animals, allowing that farmland to revert back to forest or grassland for native wildlife would be morally preferable. (The only commonly eaten farm animal actually native to North America is turkeys.)

Sandberg and Levinstein in the end do stipulate one point that may ultimately favor clean-animal products over pasture-based animal agriculture. They note that while they'd prefer more humane farming over meat breweries, the impact that farm animals have on the planet—most notably through climate-changing greenhouse gas emissions—may be sufficiently detrimental to necessitate their replacement. In this case, clean meat wouldn't be morally preferable for the farm animals themselves, but it would be preferable for humanity and the planet as a whole.

The two authors, after expressing their concerns and weighing the options, don't come down against clean meat altogether, though. "Ultimately, in vitro meat would result in great moral progress from where we currently stand, and we should continue to encourage its development. Indeed, if the option of large-scale humane farming is not possible given agricultural and economic realities, it may be our long-term best option especially given livestock's impact on the environment."

It may not, however, be a choice between clean meat and meat from happier animals—it might end up being both. By essentially all accounts, we can't have anywhere near the level of meat consumption we have today without creating vast numbers of suffering animals and incurring a range of environmental and public health costs. In order to end factory farming, we must raise fewer animals. They simply can't be treated decently when we're raising billions upon billions of them annually.

So it's possible to envision a world in which clean meat replaces a large portion of conventional animal meat but not all of it. In that world, some people may still enjoy high levels of meat consumption (most of it produced from cells, not slaughter), but some people may want the occasional conventional meat from animals who were treated well before they were killed. Just as some people still enjoy a horse-drawn carriage as a recreational experience or even for all their transportation (e.g., the Amish), some people may want meat from slaughtered animals. We'd still have some farm animals, but not factory farming. In this world, livestock may one day no longer be "live stock," but rather there'll be far fewer of them, and many simply for sentimental purposes or as companion animals. Already today, an increasing number of Americans are keeping chickens as pets, which could foreshadow more to come as we gradually shift our thinking about farm animals. And clean meat's mere existence, once commercialized, could go a long way toward shifting that thinking.

Such a debate about whether domesticated farm animals are better off existing or not may seem pretty academic to those outside the animal-ethics and cellular-ag fields. After all, nearly no one is making their food purchasing decisions based on what will maximize the sum total of happiness in the world, or even what will reduce the most amount of suffering. For better or worse, ethics are low on the list of criteria that shape most of our consumer behavior, especially when it comes to food.

Instead, survey after survey shows that what matters most when it comes to our food purchasing are three factors: price, taste, and convenience. Food sustainability advocates may wish ethics, environment, or health were competing with those stark realities, but sadly they're nowhere near the holy trinity of price, taste, and convenience.

Interestingly enough, while meat consumption fell in the United States between 2008 and 2014, hopes that the decline was due to more highly publicized concerns about sustainability may be (at least partially) unfounded. An analysis by Rabobank, a multinational food and ag-focused bank, found that American meat consumption increased by 5 percent in 2015, a huge spike, especially considering that it had been falling annually in prior years. The reason for the reversal of fortune for animal ag? "Consumers are responding to falling prices," the study's lead author asserts.

If we're waiting on more enlightened mind-sets about animals or the planet to start shifting our diets in a better direction, we might be waiting for some time. It actually may be the case that concerns about animal welfare tend to manifest once people no longer depend on those animals for their own needs. For example, once kerosene helped replace whale oil in the nineteenth century as our primary lighting fuel, it became a lot easier to start caring about the welfare of whales. Similarly, with the invention of the automobile, our view of horses became much more sentimental.

This phenomenon brings to mind the words of muckraking journalist Upton Sinclair, author of the fictionalized exposé of meat packing facilities, *The Jungle*, when he observed, "It is difficult to get a man to understand something, when his salary depends on his not understanding it." In our case, it's not that our salaries depend on exploiting animals for food (though of course some people's salaries do), but rather it's an even deeper psychological disposition. Most Americans eat meat multiple times a day and have done so their whole lives; it's firmly rooted in our culture and traditions, as it is for most humans. The very thought of becoming vegetarian, even part-time, is daunting for a lot of consumers.

Dutch philosophy professor Cor van der Weele, who's written extensively about clean meat's implications, makes the point well. "It's important to realize that change does not necessarily need to start with clear moral attitudes," she notes. "In some cases people adopt attitudes that accompany the behavior that they are already demonstrating. In this case this might mean that when people get used to eating cultured meat, the idea of factory farming or killing animals may gradually become stranger and less acceptable."

Not to imply moral equivalence, but this psychological phenomenon is often identified as one reason, for example, American states in the North became increasingly antislavery in the antebellum era, even to the point where nearly all of them succeeded in (relatively) peacefully outlawing slavery long before the Civil War. As their economies industrialized and their financial dependence on an agricultural system based on human bondage lessened, moral attitudes against slavery became more widespread in the North, at least compared to the South, which was still an almost entirely agricultural economy. At the same time, technological innovations can make a society even more dependent on an ethically abhorrent practice, too. Take the invention of the cotton gin, for example, which made the South's slave system much more lucrative for those in power. Some historians even point to Eli Whitney's invention as an inadvertent factor leading to the Civil War, since it made the South so much more resistant to legislatively ending slavery as the Northern states had mostly already done.

Thinking about such history, conservative Fox News pundit and *Washington Post* columnist Charles Krauthammer believes future generations will likely look back in horror at our treatment of animals. But he recognizes that what will lead to that may not be humane sentiment at first. Eating animals will fall by the wayside for many of us, and he says it will be "largely market-driven as well. Science will find dietary substitutes that can be produced at infinitely less cost and effort. At which point, meat will become a kind of exotic indulgence, what the cigar . . . is to the dying tobacco culture of today."

Confirming this theory, in a fascinating study looking at meat-eaters' attitudes about the mental lives of farm animals, Australian research psychologist Steve Loughnan found that, as *TIME* reported on his study, "If you like beef, you're more inclined to believe cows can't think; if you eat only fish, you're likelier to see cattle as conscious, while the salmon on your plate was probably a non-conscious nincompoop." In other words, if you eat pork, you're probably more resistant to believing studies finding that pigs are even smarter than dogs. If you eat a lot of chicken, it'll likely be hard for you to accept that chickens have language, good memories, and can even do basic math. (All true, by the way.)

Humans are great at many things, and one of them is rationalizing our conduct so we don't feel mental conflict about our behavior. As evidence shows time and again, we like to think that our behavior flows from our logically considered beliefs, but in reality, we almost always adjust our beliefs to comport with the behaviors we want to engage in. And one of those behaviors humans seem very intent to continue is eating meat.

It turns out that the maxim is true: it's easier to act your way into a new way of thinking than to think your way into a new way of acting. Once we start acting in a different way—avoiding meat from slaughtered animals—it becomes much easier to start thinking about animals in a different way, too.

Loughnan's study went on to look at whether how recently an eater had consumed meat determined his or her views about animals. Indeed, it did.

In effect, respondents were given either beef or nuts to snack on during the study, after which they were asked about the intelligence of cattle. Unsurprisingly, those who'd been given beef thought cows are far more dimwitted than did the nut consumers. It brings to mind the words of the legendary author, journalist, and animal lover Cleveland Amory, when he observed, "Man has an infinite capacity to rationalize, especially when it comes to what he wants to eat."

Amory was offering a pithy sound bite to illustrate his point, but it was hardly a new thought. In the eighteenth century, Benjamin Franklin was committed to vegetarianism on ethical grounds, but nevertheless struggled to stick with the diet and found all types of justifications to occasionally eat animals. He poignantly recalls in his autobiography:

> Hitherto I had stuck to my resolution of not eating animal food, and on this occasion consider'd, with my master Tryon, the taking every fish as a kind of unprovoked murder, since none of

them had, or ever could do us any injury that might justify the slaughter. All this seemed very reasonable. But I had formerly been a great lover of fish, and, when this came hot out of the frying-pan, it smelt admirably well. I balanc'd some time between principle and inclination, till I recollected that, when the fish were opened, I saw smaller fish taken out of their stomachs; then thought I, "If you eat one another, I don't see why we mayn't eat you." So I din'd upon cod very heartily, and continued to eat with other people, returning only now and then occasionally to a vegetable diet. So convenient a thing it is to be a reasonable creature, since it enables one to find or make a reason for everything one has a mind to do.

For years animal advocates have tried to help people see animals differently so that they'd treat them better and eat fewer of them. But what if it's the case that, for many of us, we must first eat fewer of them before we can see farm animals as individuals who matter? It really may be that as an acceptable alternative to conventional animal meat becomes more available, affordable, and frequently consumed, Americans will increasingly see farm animals as the intelligent individuals they are. As Uma Valeti of Memphis Meats predicts, "After clean meat is on the market, it's going to be unimaginable that we were okay with slaughtering billions of animals for food production despite the harm it was having to human health, the environment, and the economic inefficiencies."

———————

Our species truly is at a crossroads. It's not hard to imagine the global instability that could ensue when we have billions more people on the planet, including billions more who expect to eat meat regularly. We just don't have the resources to satiate that demand without destroying our planet and inflicting an enormity of suffering on animals, both domesticated and wild, in the process.

In the face of all the problems our seeming addiction to meat and other animal products brings, it's imperative that we initiate another green revolution of the kind that Borlaug helped foment half a century ago. Right now, among the most promising potential solutions to achieving that revolution is not to go big with animal agriculture, but rather to go small with cellular agriculture. The side effects of that revolution—increased agricultural transparency, seeing animals in a new, more respectful light, and more—would be worthy ends in themselves. But the fact that cellular ag is poised to reduce our reliance on raising animals for food is the primary reason many environmentalists, public health experts, and animal welfarists are so enthused.

The companies, nonprofit organizations, and people described in this book, and the numerous other start-ups that will assuredly form in years to come, offer a promise of the kind of increase in efficiency that we need if we're going to save the planet from ourselves. They make possible the development of a solution that could address so many of the ills our world faces, from climate change and land conservation to global hunger and animal cruelty. By producing meat and other animal products from cells and even from simple molecules, and leaving the living, breathing animals out of the equation altogether, we can achieve efficiency gains that no one in conventional animal agriculture is presently even attempting to make. We truly could bring about another green revolution.

Nations that are becoming more affluent want more meat, yet they largely lack the infrastructure and resources necessary to start a factory-farming model, let alone one that doesn't doom their environment and animals. Could cellular agriculture do for them what cellular phones did just a decade ago? Rather than developing the infrastructure needed for a first-world-type landline system connecting all their homes and businesses, many less developed nations simply leapfrogged landlines and went straight to cell phones. Already, the technology is being created to do just that for localized clean-meat production. It's not difficult to envision local meat breweries popping up in nations that might have erected factory farms instead. In a very real way, cellular agriculture could assist these nations in their move toward first-world diets.

Countries like India, a nation with one of the fastest-growing rates of meat consumption and the birthplace of Jason Matheny's quest to create a new type of meat industry, will perhaps benefit the most from such a new green revolution. Already, Indian government officials are touting the companies and people in this book by name as game changers. Cabinet minister Maneka Gandhi, after noting how many people of Indian descent are running and working at these cellular-ag companies, proudly proclaimed in 2015 that, "I have asked my niece who studies in Berkeley to intern with these companies so that one day she can look back, when the world has changed and animals are no longer killed and eaten, and be satisfied that she was a teeny part of the process. . . . Years later," she said of those who founded these start-ups, "they will be as famous as Bill Gates and Steve Jobs—that I guarantee."

New technologies have the power to radically alter our way of life, even leaving entire industries in their wake. Abraham Gesner's patenting of kerosene helped spell doom for the American whaling empire. Henry Ford's internal combustion engine rendered horse-drawn carriages archaic. It's still too early to tell just how successful the companies in this book and their competitors will be. But as they race toward the market, it's increasingly clear that cellular agriculture is no longer just a theory. It's no longer merely a prediction of Winston Churchill or Pierre-Eugène-Marcellin Berthelot. It's real, the products already exist, people (myself included) have touched and eaten them, and they may become available to consumers within a matter of years, not decades.

Does this mean a future in which antibiotics are reserved primarily for human medicine rather than as a customary animal feed additive? In which meat is far freer from dangerous bacterial contamination? In which animal agriculture causes a small a fraction of the environmental harm it does today? In which pastures and enormous fields of corn and soy are returned to forest and wetlands? In which slaughter plants give way to meat breweries? Will we soon be able to enjoy meat, eggs, milk, and leather without the twinge of guilt that plagues many of us today when we actually think about the lives and deaths of the animals destined to be our food and clothing?

Such a future may seem utopian, and there are many hurdles to making it a reality. From cost, potential regulations, consumer acceptance, and technological barriers, there's no shortage of ways that this future may be obstructed. The success of the clean-animal product movement will be far from self-executing.

But the very first steps toward that success are now clearly being taken. After all, in 2017, Uma Valeti notes that he's gotten the cost of production down more than hundredfold since Memphis Meats was founded. "We might initially enter the market at a slight price premium, but as we scale up, we are confident we will be able to produce meat at a price that is cost-competitive with (and eventually more affordable than) conventionally produced meat."

Already some of us are keeping warm with winter coats produced with lab-produced spider silk. Will we next be eating animal-free real milk yogurts, perhaps while wearing shoes of slaughter-free leather? At this pace, chicken nuggets and sausages don't seem like a bridge too far to cross.

The companies in the new field of cellular agriculture all ultimately have similar goals, but they're approaching the problem of animal agribusiness in different ways. Each believes that its particular focus is important and promising. Each shares the vision of using cellular agriculture as a means of efficiently, sustainably, and humanely helping to provide for a growing population. Their goal of a world where our meat and other animal products are produced without actual animals is an ambitious vision, and one that will take an immense amount of resources to achieve. But those resources are paltry compared to the resources the actualization of such a future would conserve.

The seemingly intractable problem of factory farming of animals has taken a heavy toll on the planet. With our ever-growing population, it's increasingly clear that it's simply no longer possible to feed all of us using such an inefficient method of production. But rather than relying on our species to change simply because it's the right thing to do, cellular agriculture may indeed prove to be the kind of change American inventor Buckminster Fuller was referencing when he declared his axiom: "To change something, build a new model that makes the existing model obsolete."

ACKNOWLEDGMENTS

My aspiration in writing this book was to do some amount of good in the world by helping familiarize readers with a nascent industry that has the potential to solve global problems and create a more humane society. Perhaps you're a reader who'll now want to get involved by joining the companies discussed or others in the field, or even by starting your own cellular-ag venture. Or maybe you're an investor and will now think about directing your attention (and of course your money) to the start-ups pioneering this new field. Or maybe you're someone who was just wondering what's the deal with slaughter-free meat and now wants to try some yourself. Or perhaps you're totally unconvinced and would never allow a cell-ag product to pass your lips. Regardless, I'm grateful to you, the reader, surely for spending your money on the book (and buying more copies for friends . . .), but more importantly for spending your valuable time reading it. Thank you.

As with any large endeavor, this book was the result of many people, all of whom have my deep gratitude. While I'd been intensely interested in the topic since the early 2000s, it was my friend Kenny Torrella who, in 2016, suggested that I write a book on it. And there's no doubt it would never have happened without my stellar agent, an Eagles fan who I call the Randall Cunningham of book agents but is known to the rest of the world as Anthony Mattero of Foundry Literary & Media. Anthony believed in this project from the first moment I mentioned it to him and was an invaluable partner from beginning to end. As he knows, the next book may be a novel about the human-animal relationship, so stay tuned!

I'm also grateful to Brooke Carey for her beneficial edits on this manuscript, which made it a stronger, and frankly, more useful book. Peter Singer, Matt Prescott, Elizabeth Castoria, Jessica Almy, and Emily Byrd offered very helpful edits to the manuscript, for which I'm grateful. I also owe a debt of gratitude to Kristie Middleton, who provided me with advice and support that was very useful. Stacy Creamer was another believer in this book whose support and guidance was very important. And of course, Adam Wilson, senior editor at Simon & Schuster's Gallery Books, was a pleasure to work with, and I'm certain he appreciated the reference to Wolverine's adamantium skeleton in the manuscript. From the moment I saw the Wolverine poster in his office, I knew we'd get along.

I'm fortunate to count many people in the field of cellular agriculture, including several who are profiled in this book, as my friends. These relationships in the industry are one of the reasons this book was possible, since they were kindly willing to share with me details about their private and often confidential work. While the book makes it clear that I'm optimistic about the promise that cellular agriculture offers to address global problems, I've done my best to maintain objectivity and report the facts as they are with as little bias as possible.

Little made me happier during the process of writing this book than Yuval Noah Harari agreeing to pen a foreword. I'm a huge fan of both of his books, *Sapiens* and *Homo Deus*, and strongly encourage you to read each of them. I learned such an immense amount from him and am still stunned to see my name on the cover of a book with his name as well. Thank you to Yuval for both the foreword and, more importantly, for helping us humans see ourselves as we are and better understand our humble place in the cosmos.

Every person, company, and organization profiled in this book was friendly, forthcoming, and a pleasure to work with. (Those who weren't didn't make it in! Just kidding.) I thank them all and hope they're content with how they're portrayed. If not, it was probably the fault of my editors, right? Sorry, Adam.

Being an author during this process wasn't my full-time job; I was working full-time at the Humane Society of the United States, too. I'm thankful to Wayne Pacelle, the president and CEO of HSUS, who was enthused about this book concept from the very beginning and encouraged me to do it. Heidi Prescott is another colleague of mine for whose support during this process I'm truly grateful. Other colleagues of mine at HSUS who reviewed the book and made useful comments on it include Rachel Querry, Bernie Unti, and Susannah May. Their edits were helpful and I'm thankful for them.

More people than I can name helped me prepare for the launch of this book, but some to whom I'm particularly grateful include Toni Okamoto and Eric Day, along with every person who blurbed this book, some of whom, like A. J. Jacobs, are among my favorite authors.

Finally, I'm grateful to my parents, Jolene and Larry Shapiro, who gave me many advantages in life and could hardly be more supportive of my work. I trust that this book will be the only gift they give to any of their friends or family again. There are also no two people more eager to eat clean meat than they are. I love them both very much!

May this book in some way help reduce the vast amount of suffering in our small world. For that, I'd be truly grateful.

ABOUT THE AUTHOR

When **PAUL SHAPIRO** took his first bite of clean meat, more humans had gone into space than had eaten real meat grown outside an animal. In addition to being among the world's first clean meat consumers, Shapiro serves as the vice president of policy engagement for the Humane Society of the United States, the world's largest animal protection organization. A TEDx speaker, founder of Compassion Over Killing, and an inductee into the Animal Rights Hall of Fame, Shapiro has published dozens of articles about animals in publications ranging from daily newspapers to academic journals. You can read more about his work and contact him at www.paul-shapiro.com.